Longman

21st Century Science

Further Science Modules

Series Editor:

Penny Johnson

Penny Johnson

Sue Kearsey

Mark Levesley

Penny Marshall

Jim Newall

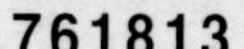

Pearson Education
Edinburgh Gate
Harlow
Essex
CM20 2JE
UK
www.longman.co.uk

First published 2007

ISBN: 978-1-4058-5558-7

Project manager and development editor: Sue Kearsey
Editors: Patrick Bonham
Design and production: Roarrdesign
Illustration: Oxford Designers & Illustrators Ltd
Picture research: Kay Altwegg
Indexer: John Holmes
Printed and bound in Great Britain by Scotprint, Haddington

The publisher's policy is to use paper manufactured from sustainable forests.

Acknowledgments

The publisher would like to thank the following for their help in reviewing this book:

Peter Borrows (Director, CLEAPSS), Phil Bradfield, Andy Piggott, Rebecca Pitkin, John Tranter (Senior Advisor, CLEAPSS) Anne Urquhart-Potts and Dr Dorothy Warren.

The publisher would like to thank the following for their kind permission to reproduce their photographs:

(Key: b-bottom; c-centre; l-left; r-right; t-top)

Action Plus: Glyn Kirk 40r, 81; Neil Tingle 38, 40t; Alamy Images: 73br, 90t; ACE STOCK LIMITED 98tl; bildagentur-online.com/th-foto / Alamy 70t; foodfolio / Alamy 70b; Roger Bamber 99b; Paul Broadbent 102; Corbis Premium Collection 51r; David Gowans 39; David R. Frazier Photolibrary 106t; foodfolio 100t; Paul Glendell 80l; Juergen Hasenkopf 53b; Image Source Pink 54l; imagebroker 105l; Cliff LeSergent 16r; Dennis MacDonald 51l; Niall McDiarmid 12; Medical-on-Line 56t; Mike Watson Images 99t; Edward Parker 104; Photofusion Picture Library 52, 103; PHOTOTAKE INC. 10b; Phototake, Inc. 46; paul ridsdale 50; Gary Roebuck 88t; Helene Rogers 98br; Shout 86b; Stockbyte Platinum 108; Thinkstock 55; Bridgeman Art Library Ltd: Academie des Sciences, Paris, France/ Archives Charmet 134r; Carnegie Institute: 136; Trevor Clifford: 24b, 62, 63b, 63c, 64, 65, 66, 69, 73l, 73tr, 76b, 76t, 82, 83, 84, 85, 88l, 89, 90bc, 90bl, 90br, 90c, 90cl, 90cr, 92l, 92r, 93, 95, 96, 97b, 97t; Corbis: Pallava Bagla 33l; Annie Griffiths Belt 34; Corbis 68; epa 151l; Jack Hollingsworth 54r; Roger Ressmeyer 146b, 147, 151r; DK Images: 19, 105r; Southern Illinois University 35l; Emerson Process Management: 31b; European Southern Observatory: 150; Food Features: Steve Moss/Food Features 67; Food Features 72; Golden Rice Humanitarian Board: 33tr; Sue Kearsey: 16l; Mark Levesley: 14, 18, 26t, 27r, 30, 32t, 33br; NASA: 113b, 149l; Centre National d'Etudes Spatiales (CNES) 112, 117l; ESA and The Hubble Heritage 135; ESA, J. Hester and A. Loll 145; ESA, S. Beckwith (STScl) and The Hubble Heritage Team 137; Goddard Space Flight Center 146t; The Hubble Heritage Team (STScl/AURA) 143l; Hui Yang University of Illinois 139; Marshall Space Flight Center 149r; SOHO 141; H. E. Bond and E. Nelan STScl 143r; naturepl.com: John Cancalosi 13; RHONDA KLEVANSKY 10t; Reinhard/ARC 24t; New Media: 74t; NHPA Ltd / Photoshot Holdings: Liinda Pitkin 41; PA Photos: Empics Sports Photograpy Ltd. 140; Pearson Education: Mari Tudor Jones 36; Photo Researchers Inc.: Sam Ogden 35r; Photodisc: 98bl, 98tr; Photofusion Picture Library: Crispin Hughes 53t; Photographers Direct: Stephen Hay Photography 24bl; Phototake, Inc: Scott Camazine 27t; Science and Society: NMPFT Daily Herald Archive 154; Science Photo Library Ltd: 57; Andrew Lambert Photography 74b, 88b, 100br; Alex Bartel 43; Biology Media 44b; Martin Bond 63t, 79; Oscar Burriel 101; Celestial Image Co. 133; CNRI 45; Martin Dohrn 8, 60; GEORGETTE DOUWMA 9b, 26bl; John Durham 31t; Dr. Fred Espenak 116t; Eye of Science 27l, 29; Robert Gendler 113t, 134l; Steve Gschemeissner 106b; TOMMASO GUICCIARDINI 78; Klaus Guldbandsen 44t; Ian Hooton 37l; Wayne Lawler 22; William Mullins 117r; J. BELL (CORNELL UNIVERSITY) / NASA 124; Faye Norman 37r; David Parker 126; Antonia Reeve 56b; J. C. Revy 9cl, 9cr, 23l, 23r; D. Roberts 48; Matthew Oldfield, Scubazoo 26br; Scott Sinklier 9t, 32bl; Eckhard Slawik 116b, 116c; Andrew Syred 28; TEK IMAGE 86t; Frank Zullo 114; Pictures Courtesy of Southern Water: 80r; STILL Pictures The Whole Earth Photo Library: Matt Meadows 32br

Cover images: Front: PunchStock: ITStock WESTEND61 Science Photo Library Ltd: Andrew Lambert Photography

All other images © Pearson Education

Every effort has been made to trace the copyright holders and we apologise in advance for any unintentional omissions. We would be pleased to insert the appropriate acknowledgement in any subsequent edition of this publication.

Edge Hill
University
Welcome to Edge Hill

Customer name: ALISON BURKE

Title: Science KS3 classbook
ID: 761831
Due: 06/10/2015 23:59

Title: Longman 21st century science
ID: 761813
Due: 06/10/2015 23:59

Total items: 2
21/09/2015 15:14

Please retain this receipt for Your Records
Edge Hill
University

Contents

Introduction to GCSE Chemistry 62

How to use this book

Each module starts with a double page that explains how the three books in the 21st Century Science series can be used to work towards a GCSE in Biology, Chemistry and Physics. There is also information about Pre-release questions, Ideas about Science and information about what you will learn in the module.

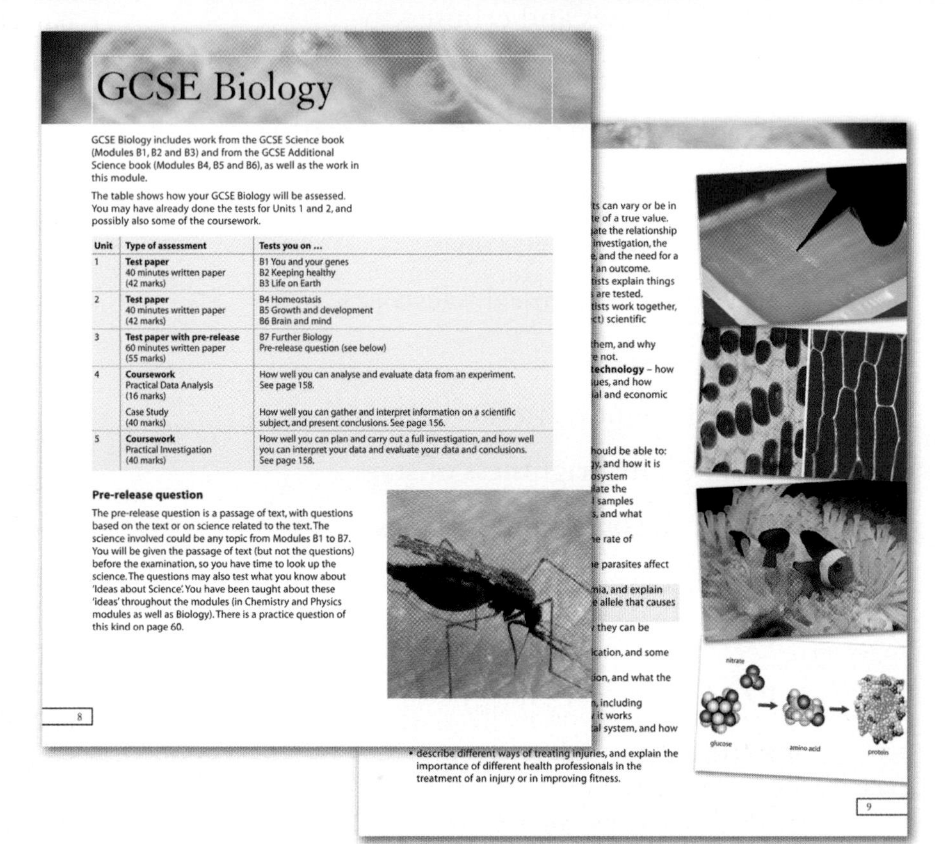

GCSE Biology

GCSE Biology includes work from the GCSE Science book (Modules B1, B2 and B3) and from the GCSE Additional Science book (Modules B4, B5 and B6), as well as the work in this module.

The table shows how your GCSE Biology will be assessed. You may have already done the tests for Units 1 and 2, and possibly also some of the coursework.

Unit	Type of assessment	Tests you on ...
1	**Test paper** 40 minutes written paper (42 marks)	B1 You and your genes B2 Keeping healthy B3 Life on Earth
2	**Test paper** 40 minutes written paper (42 marks)	B4 Homeostasis B5 Growth and development B6 Brain and mind
3	**Test paper with pre-release** 60 minutes written paper (55 marks)	B7 Further Biology Pre-release question (see below)
4	**Coursework** Practical Data Analysis (16 marks)	How well you can analyse and evaluate data from an experiment. See page 158.
	Case Study (40 marks)	How well you can gather and interpret information on a scientific subject, and present conclusions. See page 156.
5	**Coursework** Practical Investigation (40 marks)	How well you can plan and carry out a full investigation, and how well you can interpret your data and evaluate your data and conclusions. See page 158.

Pre-release question

The pre-release question is a passage of text, with questions based on the text or on science related to the text. The science involved could be any topic from Modules B1 to B7. You will be given the passage of text (but not the questions) before the examination, so you have time to look up the science. The questions may also test what you know about 'Ideas about Science'. You have been taught about these 'ideas' throughout the modules (in Chemistry and Physics modules as well as Biology). There is a practice question of this kind on page 60.

8

• describe different ways of treating injuries, and explain the importance of different health professionals in the treatment of an injury or in improving fitness.

9

Each module is divided into between 20 and 25 topics. Each topic is a double page.

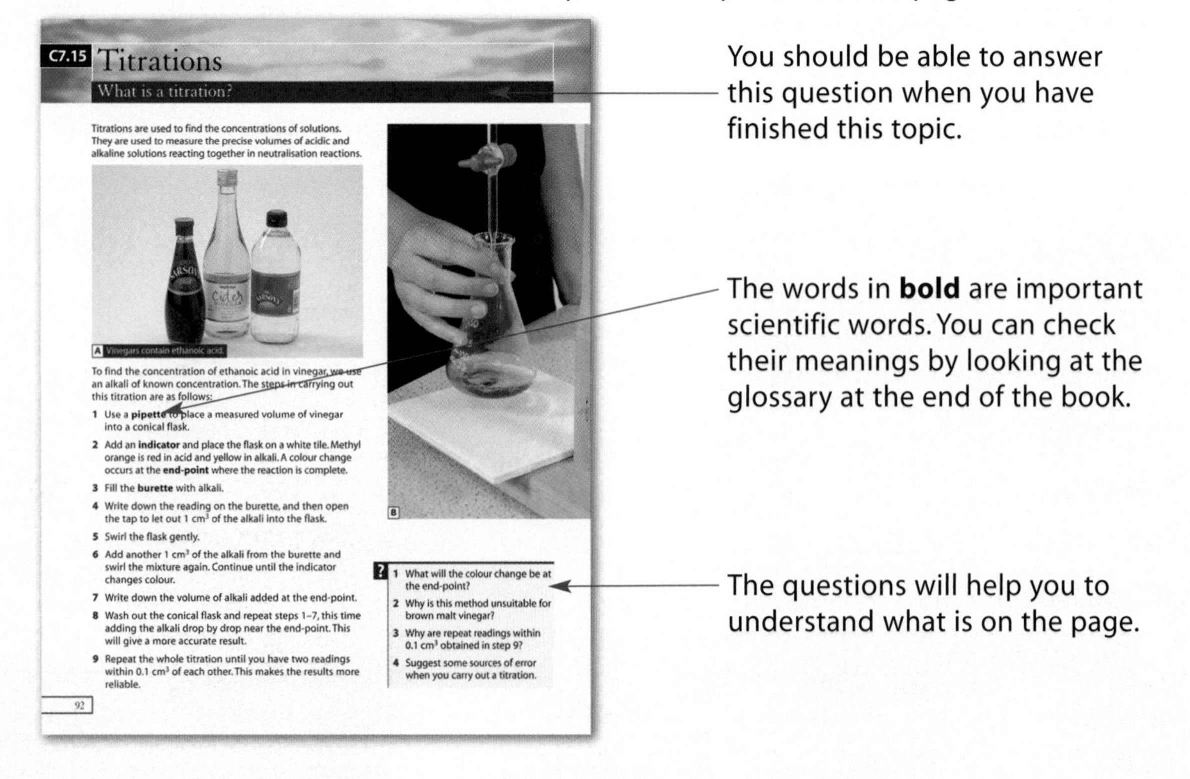

C7.15 Titrations

What is a titration?

Titrations are used to find the concentrations of solutions. They are used to measure the precise volumes of acidic and alkaline solutions reacting together in neutralisation reactions.

A Vinegars contain ethanoic acid.

To find the concentration of ethanoic acid in vinegar, we use an alkali of known concentration. The steps in carrying out this titration are as follows:

1. Use a **pipette** to place a measured volume of vinegar into a conical flask.
2. Add an **indicator** and place the flask on a white tile. Methyl orange is red in acid and yellow in alkali. A colour change occurs at the **end-point** where the reaction is complete.
3. Fill the **burette** with alkali.
4. Write down the reading on the burette, and then open the tap to let out 1 cm^3 of the alkali into the flask.
5. Swirl the flask gently.
6. Add another 1 cm^3 of the alkali from the burette and swirl the mixture again. Continue until the indicator changes colour.
7. Write down the volume of alkali added at the end-point.
8. Wash out the conical flask and repeat steps 1–7, this time adding the alkali drop by drop near the end-point. This will give a more accurate result.
9. Repeat the whole titration until you have two readings within 0.1 cm^3 of each other. This makes the results more reliable.

B

?

1. What will the colour change be at the end-point?
2. Why is this method unsuitable for brown malt vinegar?
3. Why are repeat readings within 0.1 cm^3 obtained in step 9?
4. Suggest some sources of error when you carry out a titration.

92

You should be able to answer this question when you have finished this topic.

The words in **bold** are important scientific words. You can check their meanings by looking at the glossary at the end of the book.

The questions will help you to understand what is on the page.

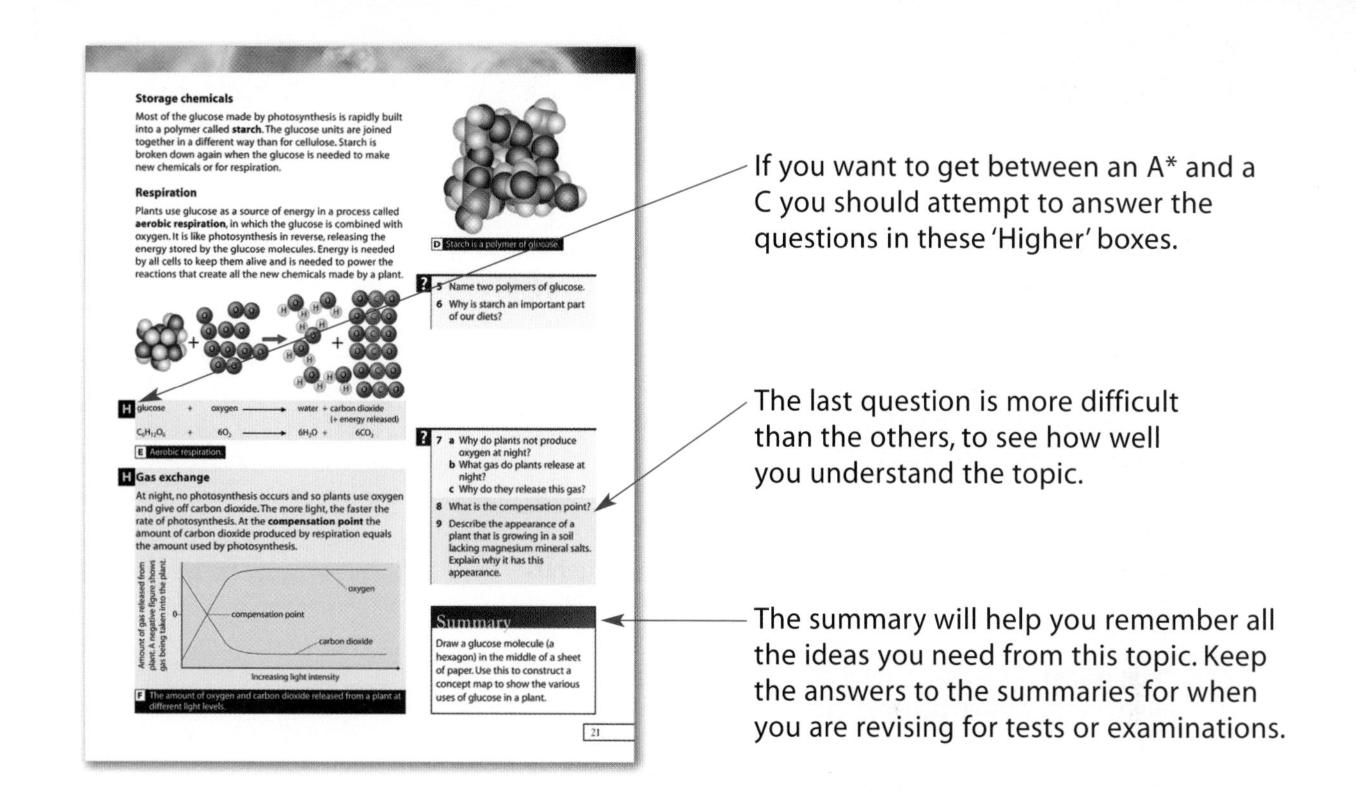

Storage chemicals

Most of the glucose made by photosynthesis is rapidly built into a polymer called **starch**. The glucose units are joined together in a different way than for cellulose. Starch is broken down again when the glucose is needed to make new chemicals or for respiration.

Respiration

Plants use glucose as a source of energy in a process called **aerobic respiration**, in which the glucose is combined with oxygen. It is like photosynthesis in reverse, releasing the energy stored by the glucose molecules. Energy is needed by all cells to keep them alive and is needed to power the reactions that create all the new chemicals made by a plant.

glucose + oxygen ⟶ water + carbon dioxide (+ energy released)

$C_6H_{12}O_6 + 6O_2 \longrightarrow 6H_2O + 6CO_2$

E Aerobic respiration.

D Starch is a polymer of glucose.

5 Name two polymers of glucose.
6 Why is starch an important part of our diets?

H **Gas exchange**

At night, no photosynthesis occurs and so plants use oxygen and give off carbon dioxide. The more light, the faster the rate of photosynthesis. At the **compensation point** the amount of carbon dioxide produced by respiration equals the amount used by photosynthesis.

F The amount of oxygen and carbon dioxide released from a plant at different light levels.

7 a Why do plants not produce oxygen at night?
b What gas do plants release at night?
c Why do they release this gas?
8 What is the compensation point?
9 Describe the appearance of a plant that is growing in a soil lacking magnesium mineral salts. Explain why it has this appearance.

Summary

Draw a glucose molecule (a hexagon) in the middle of a sheet of paper. Use this to construct a concept map to show the various uses of glucose in a plant.

21

If you want to get between an A* and a C you should attempt to answer the questions in these 'Higher' boxes.

The last question is more difficult than the others, to see how well you understand the topic.

The summary will help you remember all the ideas you need from this topic. Keep the answers to the summaries for when you are revising for tests or examinations.

There are revision questions at the end of each topic that will help you check how much you have remembered from the module. The last two pages of each module give you extra help for your exams with an example pre-release question, so you know what types of questions you might have to answer.

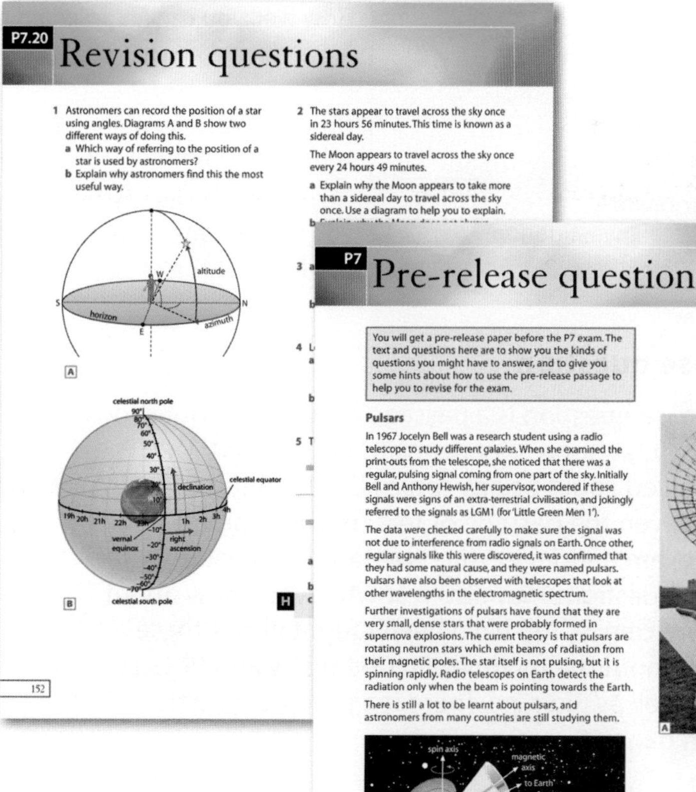

P7.20 **Revision questions**

1 Astronomers can record the position of a star using angles. Diagrams A and B show two different ways of doing this.
a Which way of referring to the position of a star is used by astronomers?
b Explain why astronomers find this the most useful way.

2 The stars appear to travel across the sky once in 23 hours 56 minutes. This time is known as a sidereal day.

The Moon appears to travel across the sky once every 24 hours 49 minutes.

a Explain why the Moon appears to take more than a sidereal day to travel across the sky once. Use a diagram to help you to explain.

152

P7 **Pre-release question**

You will get a pre-release paper before the P7 exam. The text and questions here are to show you the kinds of questions you might have to answer, and to give you some hints about how to use the pre-release passage to help you to revise for the exam.

Pulsars

In 1967 Jocelyn Bell was a research student using a radio telescope to study different galaxies. When she examined the print-outs from the telescope, she noticed that there was a regular, pulsing signal coming from one part of the sky. Initially Bell and Anthony Hewish, her supervisor, wondered if these signals were signs of an extra-terrestrial civilisation, and jokingly referred to the signals as LGM1 (for 'Little Green Men 1').

The data were checked carefully to make sure the signal was not due to interference from radio signals on Earth. Once other, regular signals like this were discovered, it was confirmed that they had some natural cause, and they were named pulsars. Pulsars have also been observed with telescopes that look at other wavelengths in the electromagnetic spectrum.

Further investigations of pulsars have found that they are very small, dense stars that were probably formed in supernova explosions. The current theory is that pulsars are rotating neutron stars which emit beams of radiation from their magnetic poles. The star itself is not pulsing, but it is spinning rapidly. Radio telescopes on Earth detect the radiation only when the beam is pointing towards the Earth.

There is still a lot to be learnt about pulsars, and astronomers from many countries are still studying them.

154

GCSE Biology

GCSE Biology includes work from the GCSE Science book (Modules B1, B2 and B3) and from the GCSE Additional Science book (Modules B4, B5 and B6), as well as the work in this module.

The table shows how your GCSE Biology will be assessed. You may have already done the tests for Units 1 and 2, and possibly also some of the coursework.

Unit	Type of assessment	Tests you on ...
1	**Test paper** 40 minutes written paper (42 marks)	B1 You and your genes B2 Keeping healthy B3 Life on Earth
2	**Test paper** 40 minutes written paper (42 marks)	B4 Homeostasis B5 Growth and development B6 Brain and mind
3	**Test paper with pre-release** 60 minutes written paper (55 marks)	B7 Further Biology Pre-release question (see below)
4	**Coursework** Practical Data Analysis (16 marks)	How well you can analyse and evaluate data from an experiment. See page 158.
	Case Study (40 marks)	How well you can gather and interpret information on a scientific subject, and present conclusions. See page 156.
5	**Coursework** Practical Investigation (40 marks)	How well you can plan and carry out a full investigation, and how well you can interpret your data and evaluate your data and conclusions. See page 158.

Pre-release question

The pre-release question is a passage of text, with questions based on the text or on science related to the text. The science involved could be any topic from Modules B1 to B7. You will be given the passage of text (but not the questions) before the examination, so you have time to look up the science. The questions may also test what you know about 'Ideas about Science'. You have been taught about these 'ideas' throughout the modules (in Chemistry and Physics modules as well as Biology). There is a practice question of this kind on page 60.

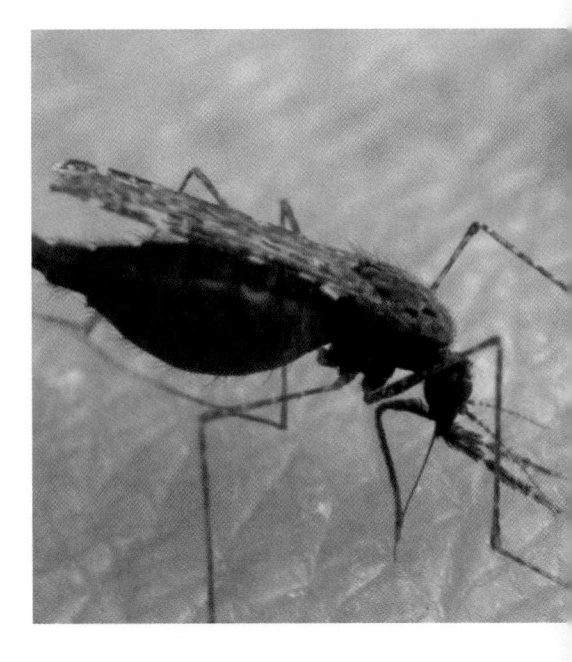

Ideas about Science

The Ideas about Science are:

- **Data and their limitations** – why results can vary or be in error, and how to make the best estimate of a true value.
- **Correlation and cause** – how to investigate the relationship between a factor and the outcome of an investigation, the difference between correlation and cause, and the need for a plausible mechanism linking a factor and an outcome.
- **Developing explanations** – how scientists explain things we observe, and how their explanations are tested.
- **The scientific community** – how scientists work together, and how they check and accept (or reject) scientific findings and explanations.
- **Risk** – what risks are, how we perceive them, and why some risks are accepted when others are not.
- **Making decisions about science and technology** – how some aspects of science raise ethical issues, and how decisions about science depend on social and economic factors as well as scientific factors.

B7 Objectives

When you have studied Module B7 you should be able to:

- describe how organisms get their energy, and how it is transferred through organisms in an ecosystem
- recall the components of soil, and calculate the percentage of water and biomass in soil samples
- recall the main stages of photosynthesis, and what happens to the glucose produced
- understand the factors that may limit the rate of photosynthesis
- explain what a parasite is, and how some parasites affect people
- H recall the symptoms of sickle-cell anaemia, and explain how natural selection has resulted in the allele that causes it to spread in certain populations
- recall the structure of bacteria, and how they can be grown on a large scale
- recall the main stages of genetic modification, and some of its uses
- describe aerobic and anaerobic respiration, and what the products of respiration are used for
- recall the parts of the circulatory system, including different blood types, and describe how it works
- describe the main features of the skeletal system, and how it can be damaged
- describe different ways of treating injuries, and explain the importance of different health professionals in the treatment of an injury or in improving fitness.

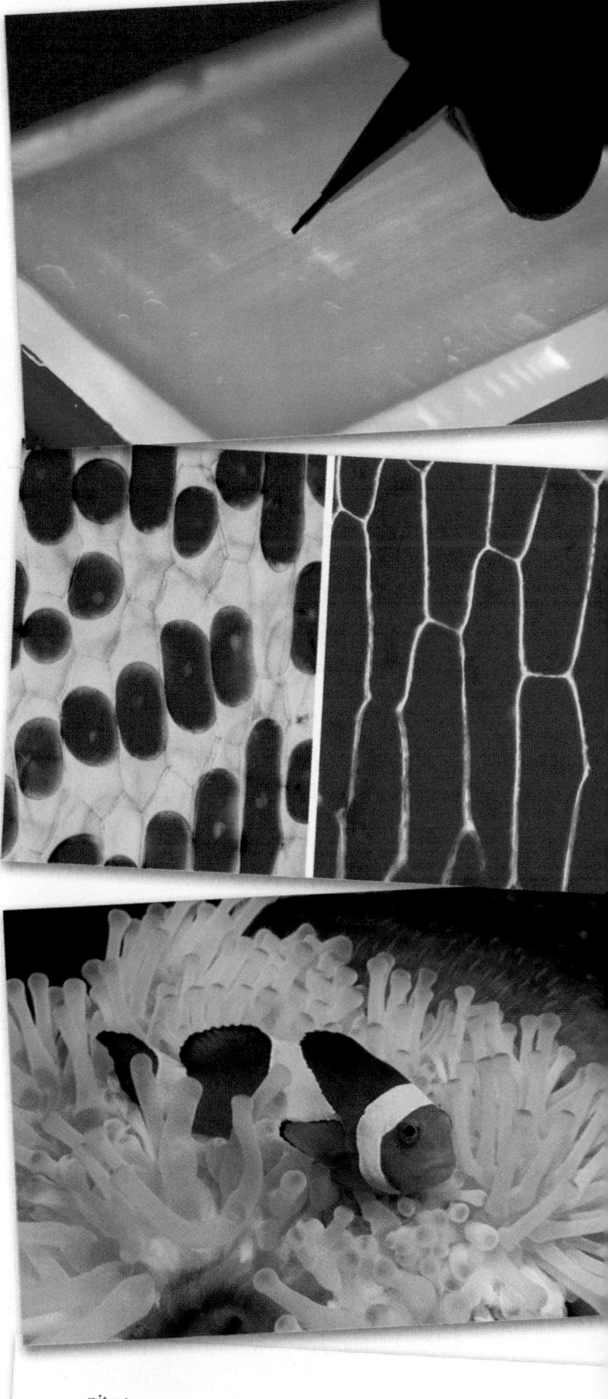

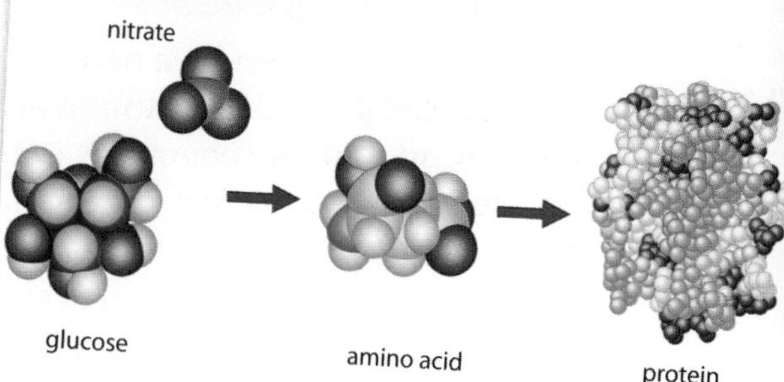

B7.1 Energy gains in food chains

How does energy enter food chains?

A Grassland is one kind of ecosystem.

An **ecosystem** contains many species of plants and animals. It also has physical conditions, such as climate and soil, that affect how the plants and animals grow. The living organisms need energy for growth, movement and all the reactions in their bodies that keep them alive.

Plants don't eat. They get their energy from food they make using the Sun's energy. Plants are **producers**. Animals eat other organisms to get energy. All animals, including humans, are **consumers**. We can use other terms for these two groups – **autotrophs** make their own food, and **heterotrophs** get their food from eating others.

The organisms in an ecosystem are linked together in **food chains**. All food chains start with an autotroph. Most autotrophs are plants that get their energy from the Sun.

Plants use light energy from the Sun to make glucose in the process of **photosynthesis**. Some of the glucose is used to build other chemicals that make up the cells of the plant. The chemicals in the plant are a store of energy.

When a plant is eaten by a heterotroph, the chemical energy stored in the plant passes to the animal. Some of this energy will be used to make chemicals in the animal's body which can then pass to another animal that eats it. A food chain shows what eats what, but it also represents the transfer of energy between the **trophic levels**.

?

3 Why can't animals, such as cows, make their own food?

4 Explain why all food chains start with an autotroph.

?

1 Write a definition for the word ecosystem.

2 a Write down the names of one heterotroph and one autotroph.

b Where does an autotroph get the energy it needs to grow?

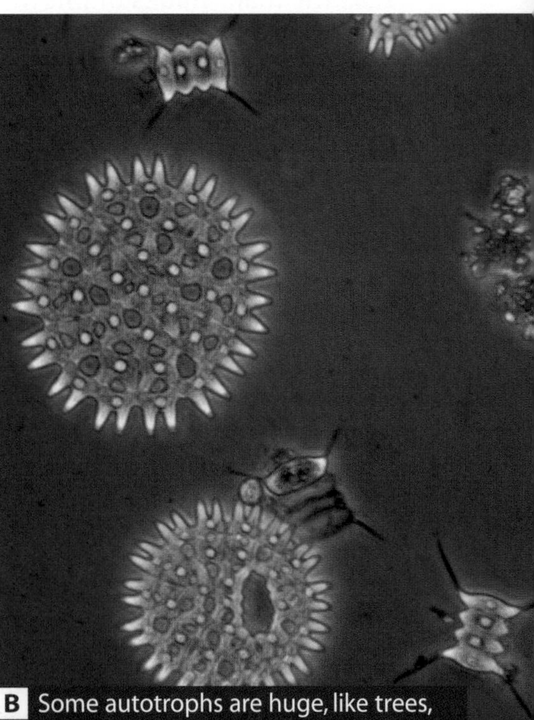

B Some autotrophs are huge, like trees, but others are microscopic, like these tiny marine plants.

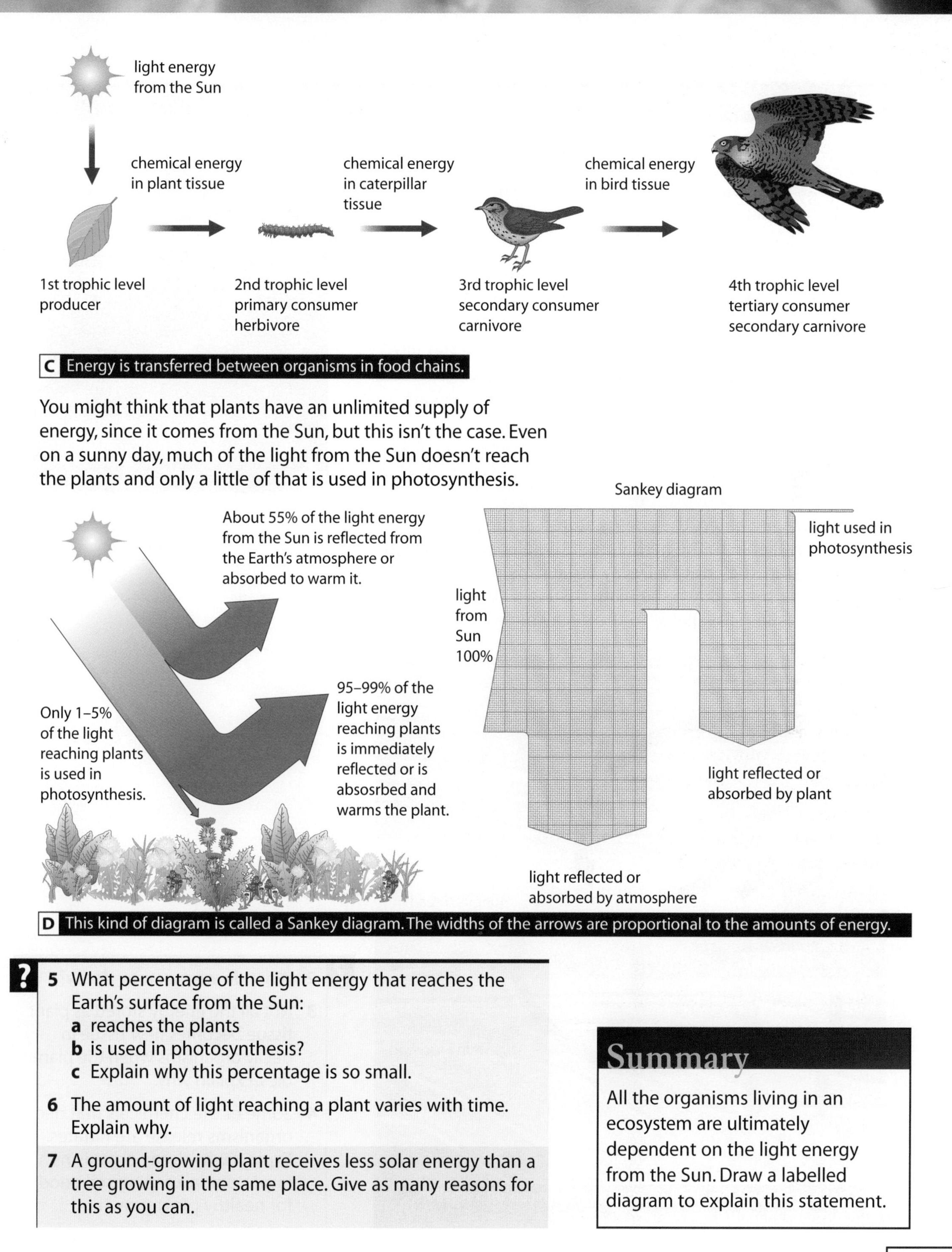

C Energy is transferred between organisms in food chains.

You might think that plants have an unlimited supply of energy, since it comes from the Sun, but this isn't the case. Even on a sunny day, much of the light from the Sun doesn't reach the plants and only a little of that is used in photosynthesis.

D This kind of diagram is called a Sankey diagram. The widths of the arrows are proportional to the amounts of energy.

?

5 What percentage of the light energy that reaches the Earth's surface from the Sun:
- **a** reaches the plants
- **b** is used in photosynthesis?
- **c** Explain why this percentage is so small.

6 The amount of light reaching a plant varies with time. Explain why.

7 A ground-growing plant receives less solar energy than a tree growing in the same place. Give as many reasons for this as you can.

Summary

All the organisms living in an ecosystem are ultimately dependent on the light energy from the Sun. Draw a labelled diagram to explain this statement.

B7.2 Energy transfers

How is energy transferred along a food chain?

The energy in a burger came from grass eaten by a cow, and the energy in the grass originally came from the Sun. But only a small amount of the energy that the grass trapped in photosynthesis ended up in the burger.

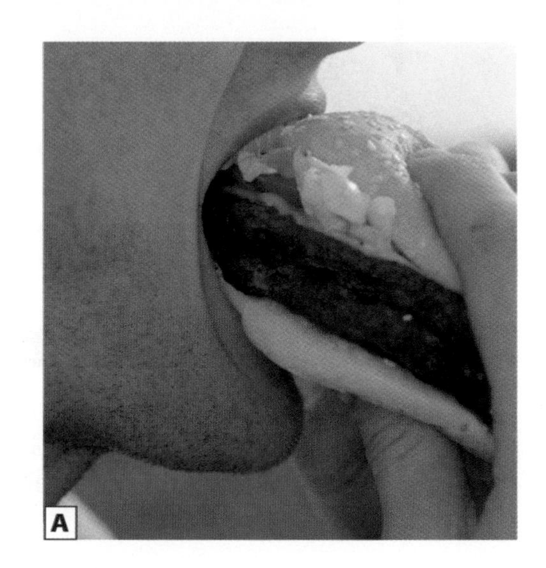

A

Most plants transfer less than 5% of the light energy they receive into chemical energy stored in plant tissues. The plants release some of this stored energy to carry out all the processes they need in order to grow, reproduce and stay alive.

?

1 **a** What is the process that releases energy from stored glucose?
b In which organisms does this process happen? Explain your answer.

When plants are eaten, the energy stored in the plant becomes available to the herbivore. If the plants die instead of being eaten, they **decay**. Decay organisms, such as fungi and bacteria, break down the plant tissues, so the energy in the plant tissues is available to the decay organisms.

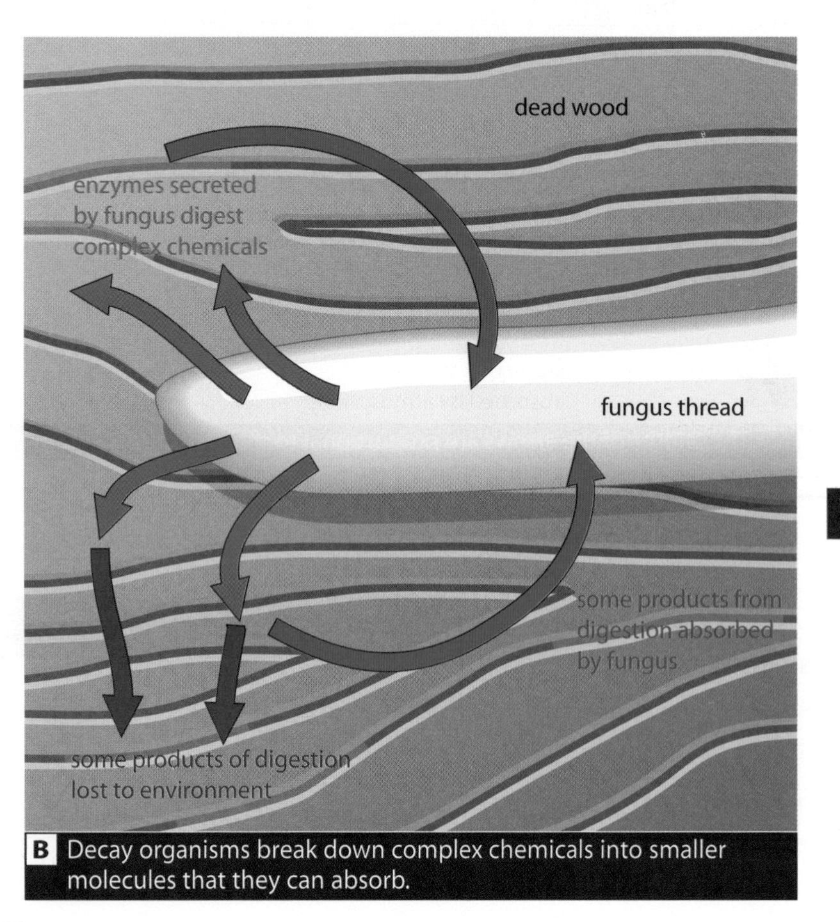

B Decay organisms break down complex chemicals into smaller molecules that they can absorb.

?

2 What do we mean by decay?

3 Not all the energy stored as plant tissue becomes new tissue in decay organisms when the plant dies. Explain why.

4 Some of the chemicals that decay organisms release are nitrates. Explain why decay organisms are important in keeping soil good for healthy plant growth.

When an animal eats a plant, not all the energy stored in the plant becomes animal tissues. Plant tissues are not easy to digest, so some will never be absorbed into the animal's body. They will be **egested** in faeces. Decay organisms can digest the complex chemicals in faeces, so they feed on this as well as on dead organisms.

C A koala can eat only the leaves from this tree. The rest of the tree's tissues are too difficult for it to digest.

?

5 Explain why a herbivore absorbs only some of the energy in its food.

When the chemicals from a herbivore's food are absorbed into its body, they won't all be converted to new animal tissues. Only the energy in those chemicals that become part of its body will be available to an animal that eats it, or to the decay organisms that feed on it after it dies.

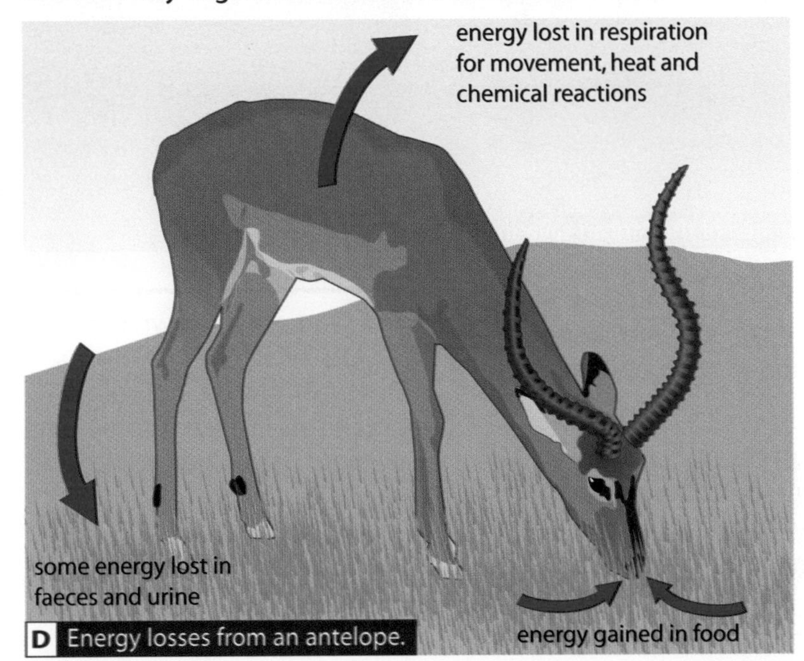

D Energy losses from an antelope.

E We can calculate the **efficiency of energy transfer** from one trophic level in a food chain to another using this equation:

$$\text{efficiency} = \frac{\text{energy in tissues of organisms at one trophic level}}{\text{energy in tissues of organisms at level below}} \times 100\%$$

The efficiency of energy transfer in diagram D is about 3%, but for other organisms it can be much more or less than this.

?

6 List all the energy losses from a herbivore, such as an antelope.

7 Use diagram D to help you draw another diagram that shows the energy losses from a carnivore, such as a lion.

8 Some of the energy losses from a crocodile are much less than those of a lion. Explain why.

Summary

The amount of energy transferred from one level to the next in a food chain gets less as you go up the chain. Draw an annotated diagram of a food chain to explain why this is.

B7.3 Feeding pyramids

What are pyramids of number and biomass?

A Why are there only a few hyenas in this huge area of the African savannah?

When you are studying the organisms in a food chain, there are lots of different kinds of data you can collect. One of the simplest things to do is to count the number of organisms at each trophic level.

Diagram B shows the information from photo A drawn as a **pyramid of numbers**. This shows the number of organisms at each trophic level in the food chain, with the producers at the bottom. A pyramid of numbers is drawn to scale, as far as possible, so it is easy to compare the differences in numbers between levels, although here the differences are too great to draw the pyramid to scale.

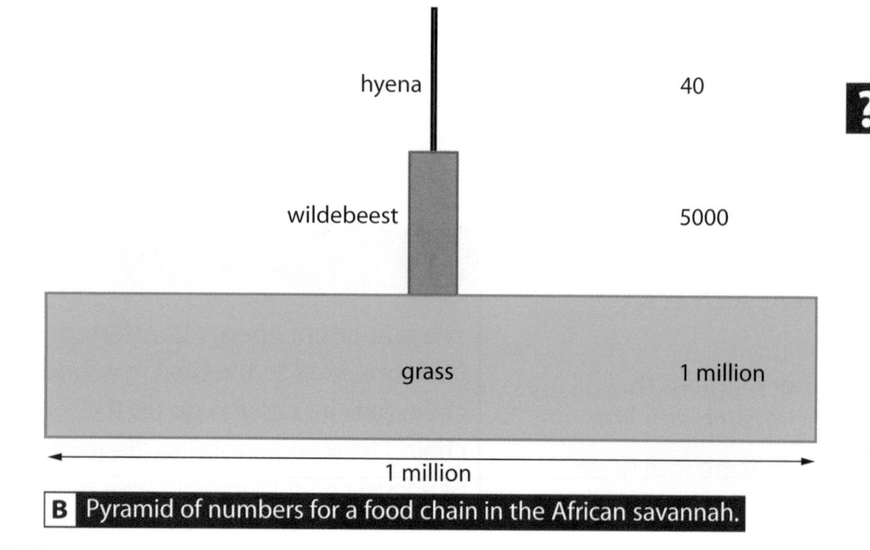

B Pyramid of numbers for a food chain in the African savannah.

?

1 The food chain shown in photo A is grass→wildebeest→hyena. Describe how the number of organisms in each trophic level changes as you go up the food chain.

?

2 a Owls eat fieldmice, which eat blackberries. In an area being studied there was one bush with 102 blackberries, 16 fieldmice and one owl. Draw a pyramid of numbers to show this information.

b From what you have learnt about energy flow through a food chain, suggest why pyramids of numbers often have this shape.

The information for a pyramid of numbers is relatively easy to collect. However, the data can often give a shape that is not a pyramid.

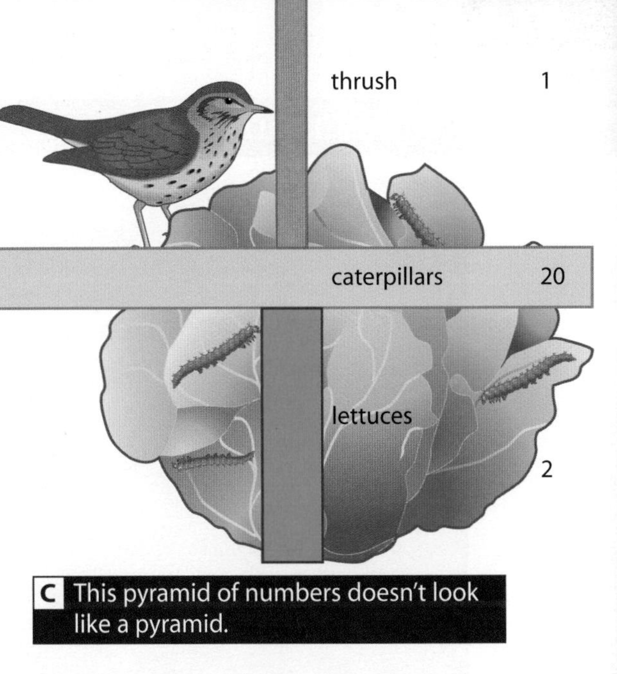

C This pyramid of numbers doesn't look like a pyramid.

?

3 Suggest why the pyramid of numbers in diagram C is not the right shape.

4 Give one advantage and one disadvantage of pyramids of numbers.

Another way to show relationships in a food chain is to convert the numbers to **biomass**, which is the mass of the organisms. The total biomass in a trophic level can be calculated by multiplying the number of organisms by the average mass of one of the organisms. This information can be displayed as a **pyramid of biomass**. This kind of pyramid is also drawn to scale and starts with the producers at the bottom level.

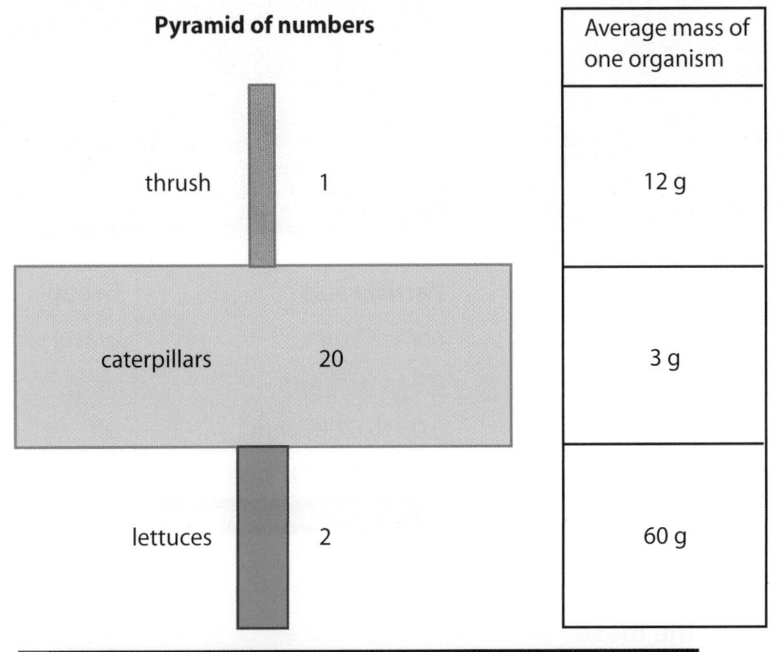

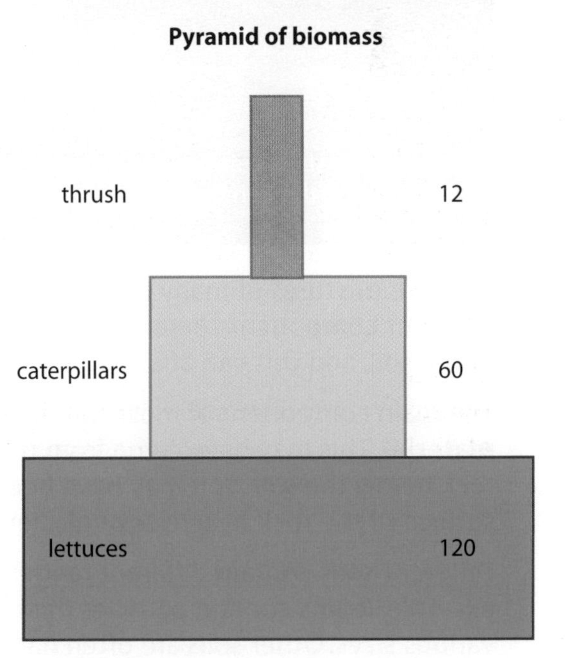

D Converting a pyramid of numbers to a pyramid of biomass.

?

5 Explain why the pyramid of numbers in diagram C converts to a pyramid of biomass that is a pyramid shape.

6 a Give one advantage of pyramids of biomass over pyramids of numbers.

b One problem with calculating a pyramid of biomass is finding an accurate estimate for the average mass of the organisms at each level of a food chain. Look at photo A and explain why this can be a problem.

7 Give one criticism of both pyramids of numbers and pyramids of biomass. (*Hint*: think of food webs rather than food chains.)

Summary

Make a poster on pyramids of numbers and biomass. Consider a suitable food chain that you could use and suggest how you would collect the data for each pyramid. Add brief notes on the advantages and disadvantages of each kind of pyramid.

B7.4 The composition of soil

What is in soil?

A Many more sheep can feed on the grass on the rich soil in the background than on the same area of grass on the poor soil in the foreground.

B A lot of gravel, sand and clay were dropped in the UK when glaciers retreated at the end of the last ice age.

Soils are mixtures of many components. The proportions of different components determine the nature and properties of the soil, and this can affect how plants grow.

The main component of most soils is the **inorganic material**. This may have come from the breaking down of rock below the soil, or it may have been left behind on the surface of the rock by a river or glacier.

Different soils contain different ranges of particle sizes. For example, loams contain particles of sand, silt and clay of various sizes. Other soils are often named after the most common particle size.

Particle size	Group
above 2 mm	gravel
0.2 to 0.06 mm	sand
0.06 to 0.002 mm	silt
less than 0.002 mm	clay

C Particle groups in soils.

?

1 Give three sources of the inorganic material in soils.

2 What would the following soils be called?
   a a soil with most particles about 0.001 mm in size
   b a soil with most particles between 0.04 and 0.1 mm.

The size of the particles affects the amount of air and water in the soil. Large particles do not fit closely together. This means there is space for air between the particles, but water drains away quickly. Very small particles fit together closely, which leaves little space for air and makes it difficult for water to soak down between them.

?

3 Which soil would dry faster after rain: a sandy soil or a clay soil? Explain your answer.

4 Clay soils can be difficult to dig after rain. Suggest why.

5 Most plants need air and water at their roots. Suggest the best soil for a garden. Explain your answer.

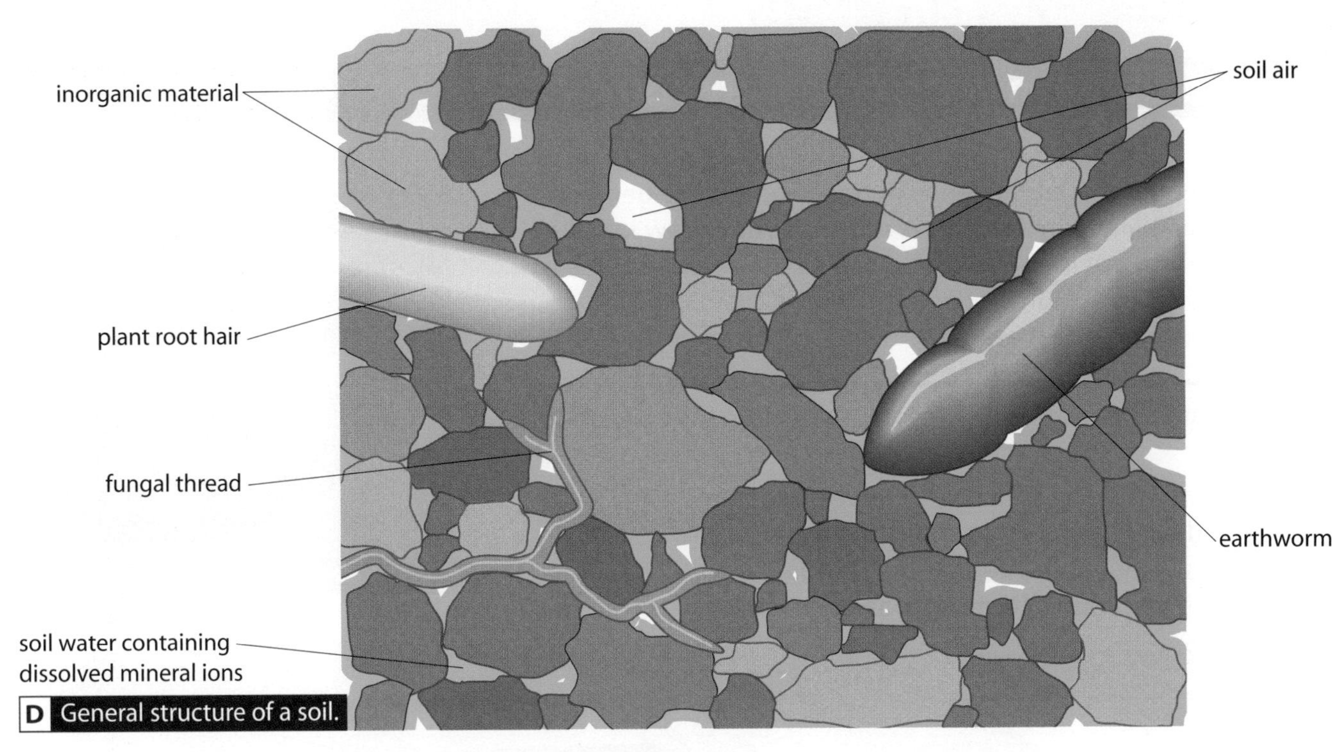

D General structure of a soil.

Soils also contain biomass. Some of this is the organisms that live in it; the rest is **organic material** from dead organisms. When this is broken down by decay organisms, it releases ions such as nitrates. These ions dissolve in soil water along with other ions, such as magnesium and calcium ions, that have come from the inorganic part of the soil. Ions are quickly **leached** (washed out) from sandy soils.

A The percentage of water in a soil can be calculated as:

$$\frac{\text{mass of water in soil}}{\text{mass of sample}} \times 100\%$$

- How would you find each of the masses?
- How would you change this to find the percentage biomass in a soil?

?

6 Many plants do not grow well in sandy soils. Explain why.

7 Gardeners often add compost or manure to their soil. Explain what happens to this in the soil, and why gardeners think this is important.

8 Peat soils are mostly organic material. They formed in boggy areas such as the fens and moors. Much of the UK's wheat grows on drained fen soils. Explain why these drained soils are so fertile, and why they are slowly becoming less fertile.

9 Explain how some soils can support food chains that contain more energy than others, even though they have the same weather conditions.

Summary

Use the information in this topic to explain the differences in the plant growth shown in photo A.

B7.5 Photosynthesis

What happens during photosynthesis?

Some people think that Venus flytraps eat food to get energy, just like humans. This is not true. They grow in soils that lack **nitrate** ions and catch insects to use as a source of nitrates, *not* as a source of energy. Like all plants, they photosynthesise to make their own food for energy.

A Venus flytrap.

?

1 What process do Venus flytraps use to make their own food?

Photosynthesis occurs mainly in leaves. It requires water and carbon dioxide as **raw materials**. The water travels up from the roots in **xylem tissue**, and the carbon dioxide **diffuses** into the leaf through **stomata** (one of which is called a stoma).

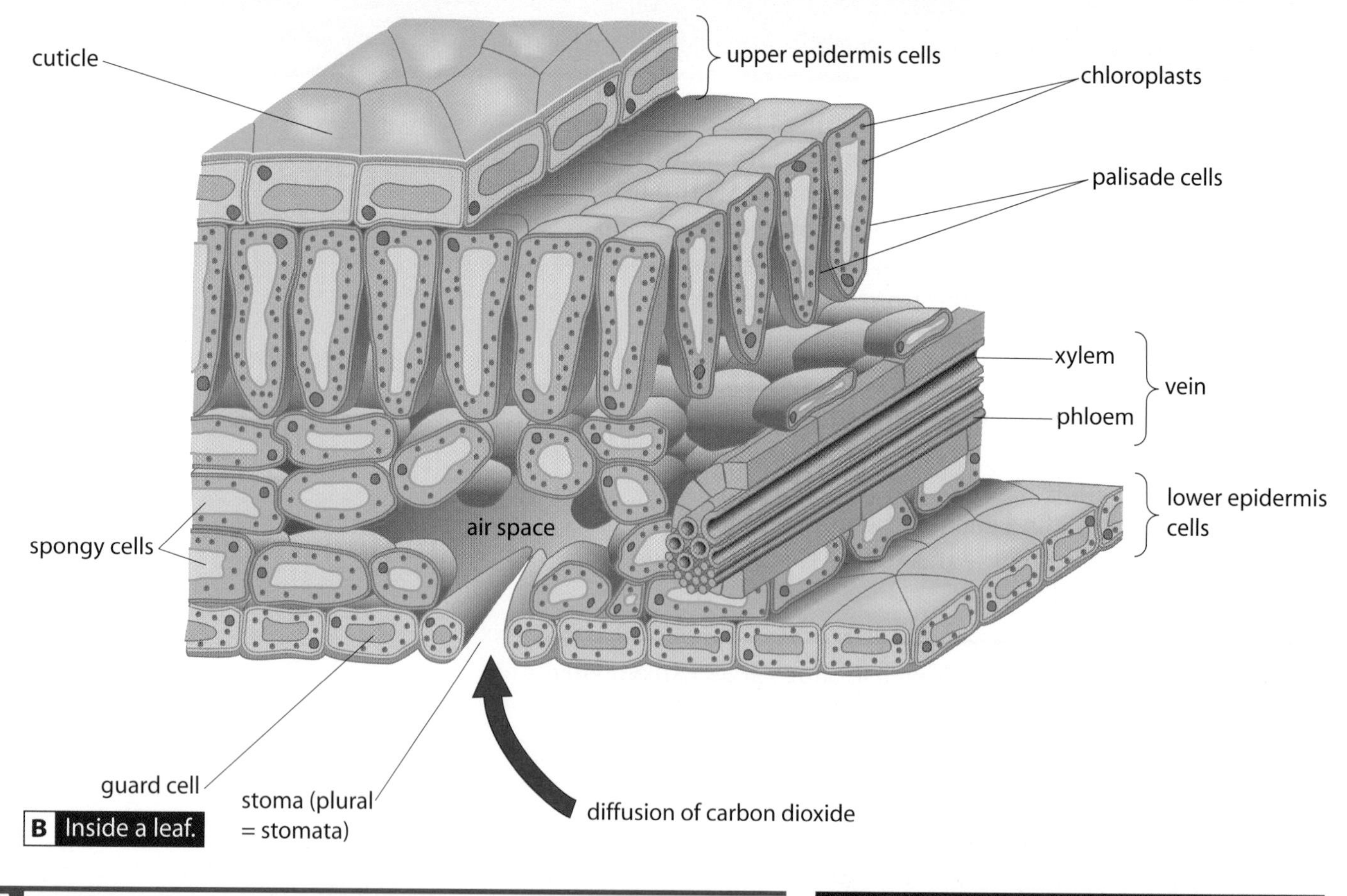

B Inside a leaf.

A

If leaves are put into very hot water, gas bubbles out of the stomata.

- How would you show where stomata are found in different leaves?
- Are stomata found in different places depending on where a plant lives?

?

2 What are the raw materials for photosynthesis?

3 In what tissue is water carried up a plant?

4 What is a stoma?

Chloroplasts and chlorophyll

Photosynthesis occurs in **chloroplasts** – disc-shaped structures that contain a green substance called **chlorophyll**. The chloroplasts can rise in the cells to increase the amount of light they receive, or sink to protect themselves from damage by very bright light.

C Inside a chloroplast. In 1967 Prof. Lynn Margulis (born 1938) proposed that a couple of billion years ago, photosynthesising bacteria got inside cells that could not photosynthesise and evolved into chloroplasts, turning the cells into the first plant cells. This is the accepted theory today.

?

5 Look at diagram B.
- **a** In which cells does most photosynthesis occur?
- **b** How can you tell?

Chlorophyll absorbs light energy and uses it to split water molecules into oxygen and hydrogen atoms. The oxygen atoms form O_2 molecules and diffuse out of the leaf. The hydrogen atoms react with carbon dioxide to form a **sugar** called **glucose** (and more waste oxygen). Photosynthesis is a process made up of a number of different reactions – it is not a single reaction.

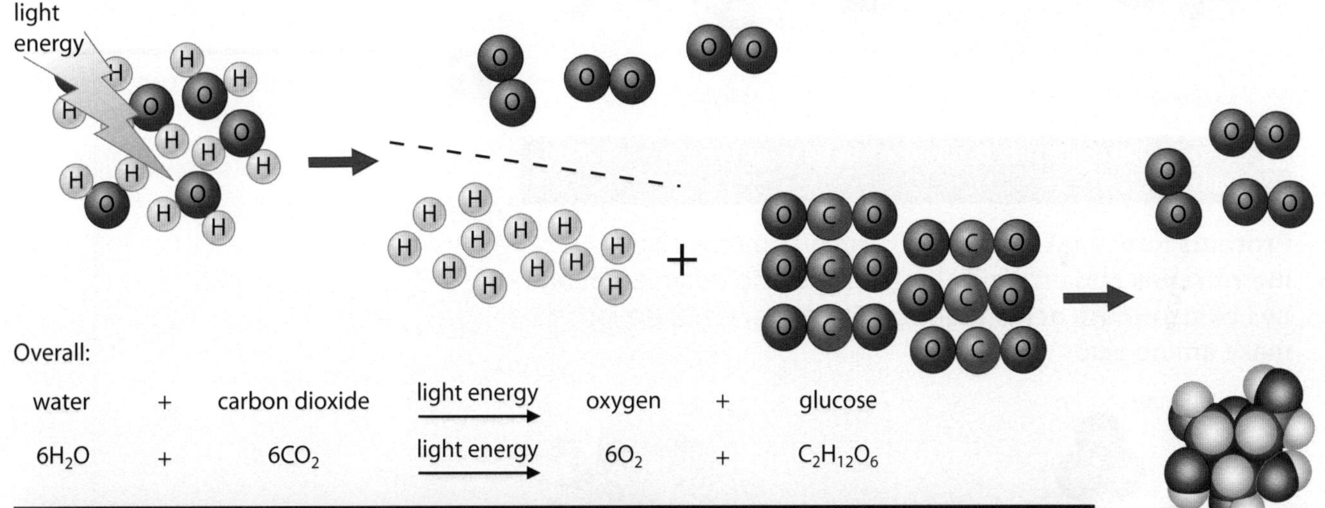

D In photosynthesis, light energy is used to help rearrange the atoms in water and carbon dioxide, to form glucose and waste oxygen.

?

6 Through which parts of a leaf do oxygen molecules diffuse out of a plant?

7 For every six molecules of water and six molecules of carbon dioxide, how many molecules of glucose are produced in photosynthesis?

8 Why is oxygen not produced by plants when it is dark?

9 Chloroplasts can move up and down in cells. Why do you think they do this?

10 Water lilies have leaves that float on water. Where would you expect to find stomata on a water lily leaf? Explain your answer.

Summary

Write out the word equation for photosynthesis and label it with notes to summarise these two pages. Include all the words in bold.

B7.6 Uses of glucose

What is the glucose made in photosynthesis used for?

A large tree produces about 300 g of glucose on a sunny day using photosynthesis. This glucose is then used to make all the chemicals found in plants.

Chemicals for growth

In order to grow, a plant must produce new substances with which to build new cells. **Cellulose** is made by joining glucose molecules together to form long **polymer** chains. Cellulose chains are then used to build **cell walls**, which provide support.

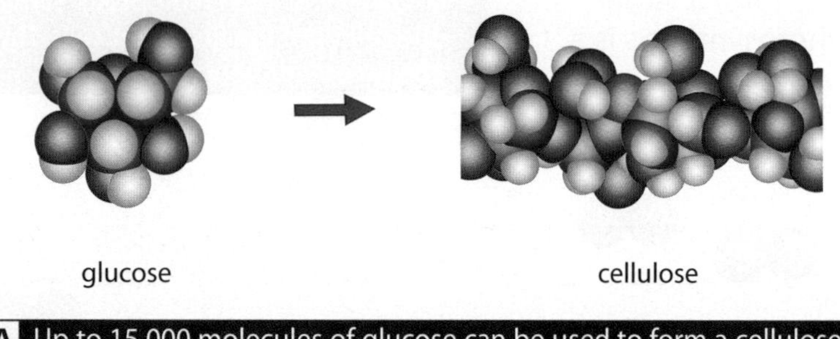

A Up to 15 000 molecules of glucose can be used to form a cellulose chain.

?

1 What are the two products of photosynthesis?

2 Why do plants make cellulose?

Proteins form vital parts of cells, such as parts of **cell membranes** and enzymes. Proteins are also polymers, made by linking **amino acids** together. Plants need nitrate ions to make amino acids.

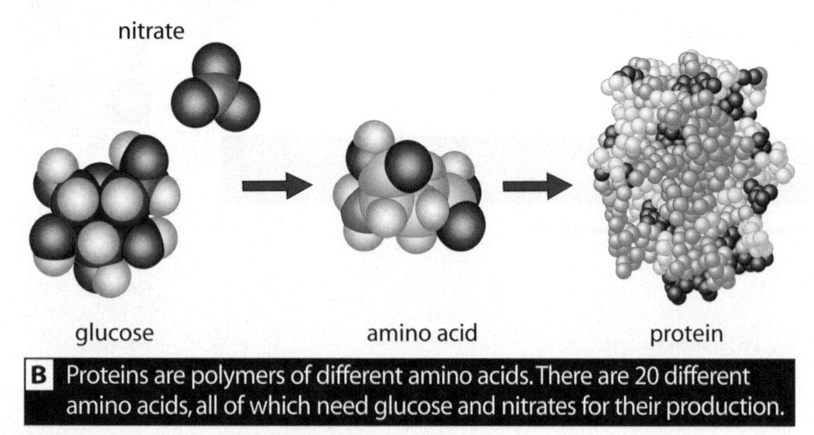

B Proteins are polymers of different amino acids. There are 20 different amino acids, all of which need glucose and nitrates for their production.

Mineral salts containing magnesium ions are needed to make chlorophyll molecules.

?

3 **a** Plants in soils with few nitrates grow poorly. Why?
b Name a plant that can grow well in nitrate-poor soils.

4 Explain why chlorophyll is not described as a polymer.

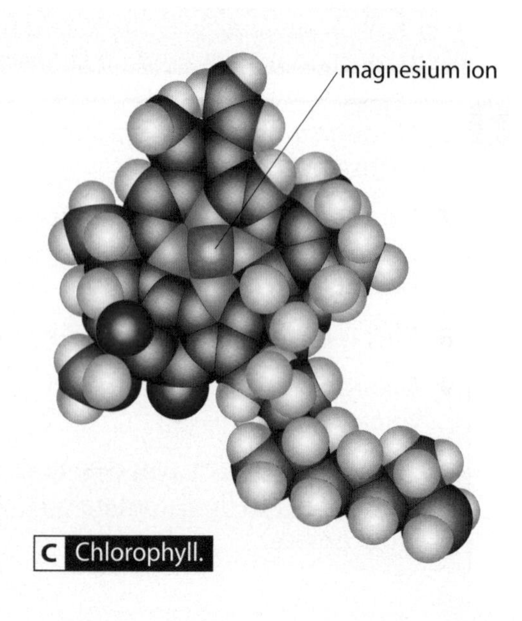

C Chlorophyll.

Storage chemicals

Most of the glucose made by photosynthesis is rapidly built into a polymer called **starch**. The glucose units are joined together in a different way than for cellulose. Starch is broken down again when the glucose is needed to make new chemicals or for respiration.

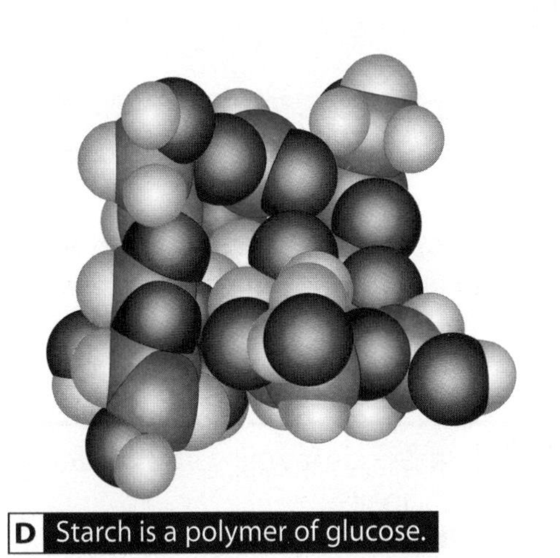

D Starch is a polymer of glucose.

Respiration

Plants use glucose as a source of energy in a process called **aerobic respiration**, in which the glucose is combined with oxygen. It is like photosynthesis in reverse, releasing the energy stored by the glucose molecules. Energy is needed by all cells to keep them alive and is needed to power the reactions that create all the new chemicals made by a plant.

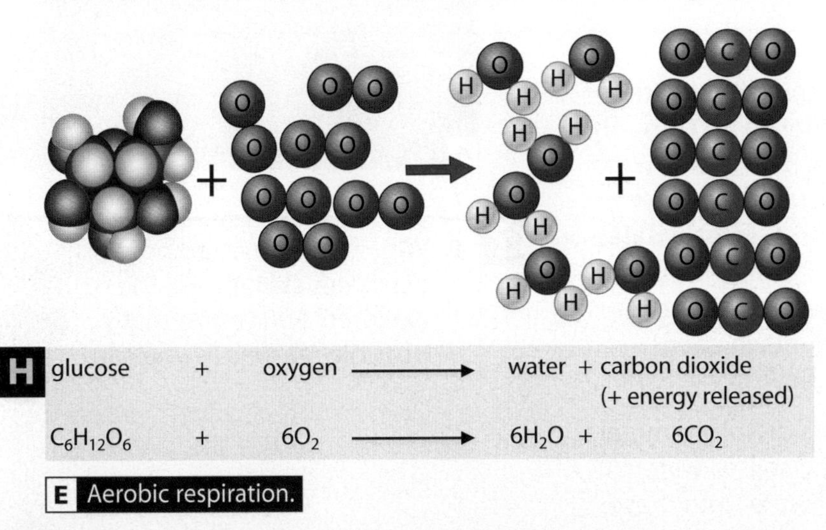

H glucose + oxygen → water + carbon dioxide (+ energy released)

$C_6H_{12}O_6 + 6O_2 \longrightarrow 6H_2O + 6CO_2$

E Aerobic respiration.

> **?**
> 5 Name two polymers of glucose.
> 6 Why is starch an important part of our diets?

H Gas exchange

At night, no photosynthesis occurs and so plants use oxygen and give off carbon dioxide. The more light, the faster the rate of photosynthesis. At the **compensation point** the amount of carbon dioxide produced by respiration equals the amount used by photosynthesis.

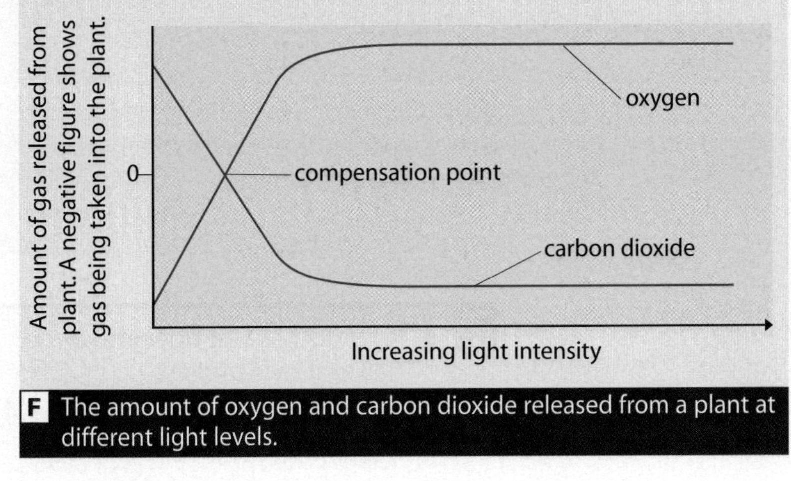

F The amount of oxygen and carbon dioxide released from a plant at different light levels.

> **?**
> 7 **a** Why do plants not produce oxygen at night?
> **b** What gas do plants release at night?
> **c** Why do they release this gas?
> 8 What is the compensation point?
> 9 Describe the appearance of a plant that is growing in a soil lacking magnesium mineral salts. Explain why it has this appearance.

Summary

Draw a glucose molecule (a hexagon) in the middle of a sheet of paper. Use this to construct a concept map to show the various uses of glucose in a plant.

B7.7 Water and nitrate uptake

How are water and mineral ions taken into a plant?

H Explorers often rub salt onto exposed skin when visiting leech-infested areas. Leeches have thin skins and contain lots of water, and the salt causes the water to flow out of their bodies by **osmosis**, causing death by dehydration.

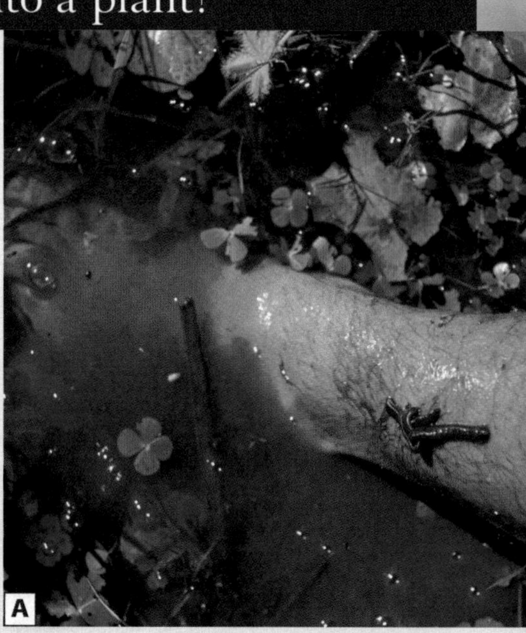
A

?

1. **a** Suggest why some gardeners put salt on slugs and snails.
   b Suggest why gardeners don't just spread salt all over their flowerbeds.

?

2. Why will salt rubbed onto an explorer's skin not kill the explorer? Suggest as many possible reasons as you can.

Osmosis

A **partially permeable membrane** is a thin sheet with holes that allow small molecules like water to go through, but not bigger molecules. Osmosis is the net flow of water from a dilute solution to a more concentrated solution through a partially permeable membrane. **Osmotic balance** is achieved when the concentrations on both sides of the membrane are equal. Water molecules will still move through the membrane, but the movements in each direction will be equal – there is no net flow.

The **cytoplasm** of **root hair cells** contains dissolved mineral ions at a greater concentration than in the soil, so water enters these cells by osmosis. The flow of water into these cells pushes water towards the xylem to be carried up the plant. Root hair cells have ways to make sure that osmotic balance is never achieved, and the cytoplasm always contains a greater concentration of mineral ions than the soil water.

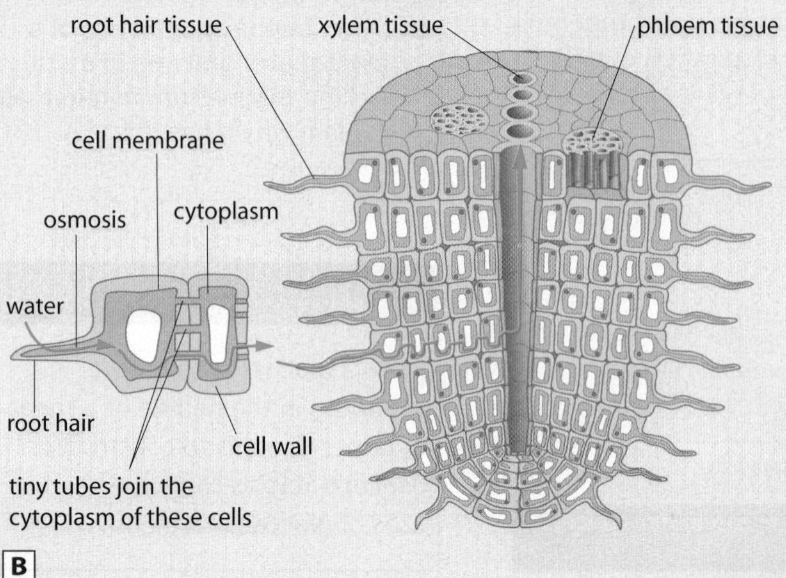

B

?

3. Why does water flow into a root hair cell by osmosis?
4. What is osmotic balance?

In a leaf, water diffuses out of the xylem. Some moves into the cells by osmosis, because the cytoplasm contains a more concentrated solution of solutes. In a cell that is photosynthesising, glucose is produced. Making an increasingly strong glucose solution inside a cell would cause more and more water to flow into it. To avoid this, the glucose is converted into insoluble starch, which has little effect on the osmotic balance of the cell.

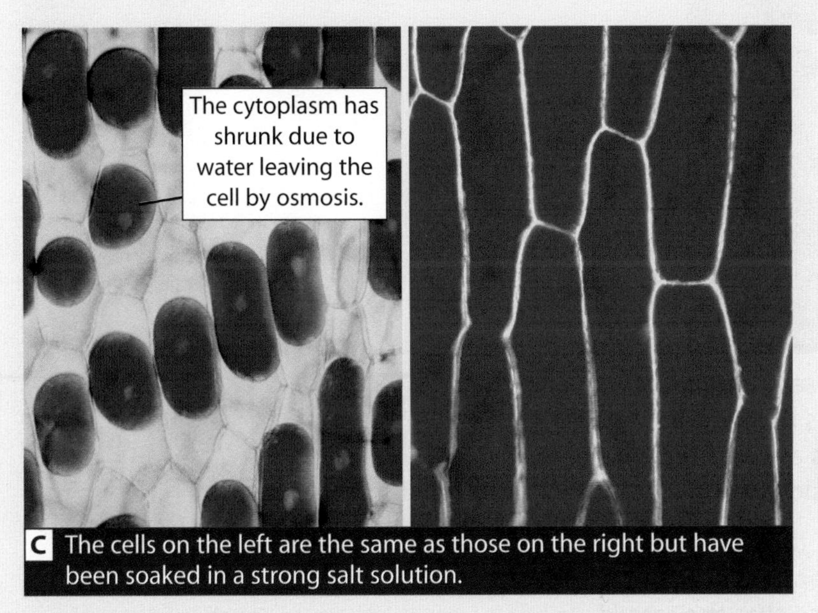

C The cells on the left are the same as those on the right but have been soaked in a strong salt solution.

?

5 What might happen if the glucose from photosynthesis were not converted into starch?

Active transport

The concentration of mineral ions in the cytoplasm of a root hair cell is up to 100 times greater than the concentration in the soil. The cell surface membranes of root hair cells maintain this imbalance. Embedded in the membranes are molecules that act as pumps, which attach to mineral ions and pull them into the cells. This process requires energy and is called **active transport**.

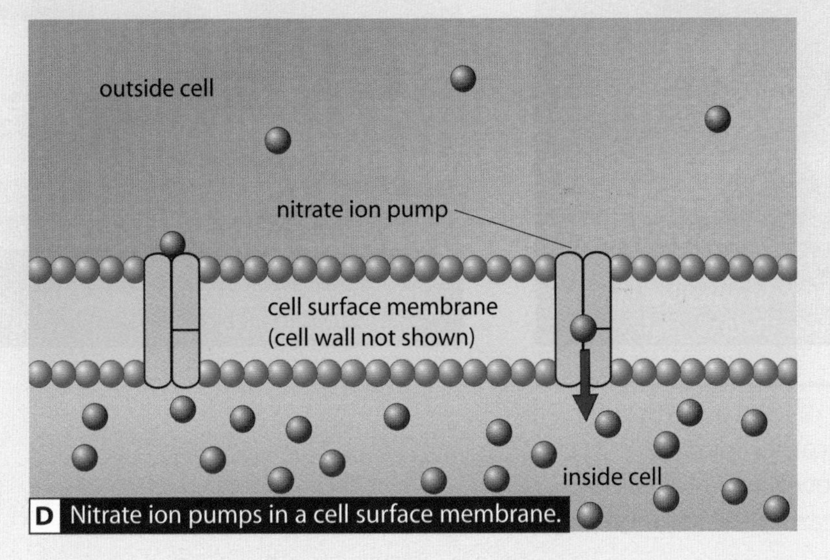

D Nitrate ion pumps in a cell surface membrane.

?

6 Explain how active transport of nitrates also helps water to enter a root hair cell.

7 Cyanide stops respiration occurring in root hair cells.
   - **a** Cyanide decreases the mineral ion concentration imbalance between cytoplasm and soil. Why does this happen?
   - **b** What other piece of evidence suggests that mineral ion uptake is an active process?

Summary

Draw a root hair cell and label it to show how water and nitrates enter it.

B7.8 Rate of photosynthesis

What factors control the rate of photosynthesis?

A Mountain plants are often slow growing and small.

In mountain regions the air is less dense; there are fewer air molecules than at sea level. This means that there is less carbon dioxide for photosynthesis. It is also colder. For these reasons, plants in these regions grow slowly.

The **rate** of photosynthesis is how fast it happens, and is affected by the amounts of heat, carbon dioxide and light. If one of these things is in short supply, so that the rate is slowed, that thing is a **limiting factor**.

The rate of photosynthesis can be measured by finding out how much glucose or oxygen is produced in a certain time.

?

1. List three factors that can limit the rate of photosynthesis.
2. Which factor is *not* limiting in mountain regions?

B Balls of this photosynthesising alga float during the day as oxygen collects inside them. Some plant leaves can be made to behave in the same way; half cress leaves can be made to sink in water, but bubbles of oxygen from photosynthesis will then cause them to float.

C Oxygen bubbles can be seen rising from some photosynthesising water plants. This plant is *Elodea*.

A How would you find out how the rate of photosynthesis is affected by light, temperature and/or carbon dioxide concentration? (*Hint*: carbon dioxide levels in water can be increased by using sodium hydrogencarbonate solution.)

Graphs can be drawn to show how some factors affect the rate of photosynthesis.

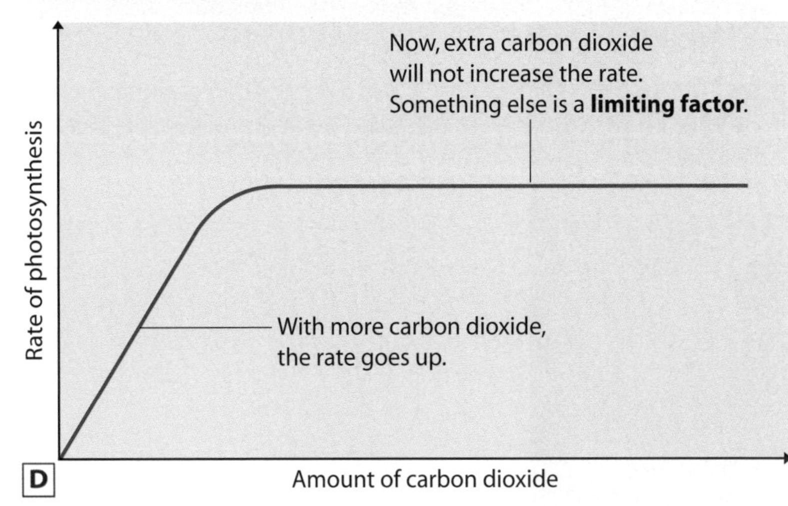

> **?** **3** Where graph D levels off, something has become a limiting factor. Which of these could it be?
> carbon dioxide glucose
> light nitrogen oxygen
> sugar temperature

Increasing the temperature increases the rate of photosynthesis, because it gives the **enzymes** and reactants more energy and allows them to work faster. However, if the temperature is too high the enzymes become **denatured** (lose their shapes) so that they no longer work.

> **?** **4** Look at graph E. At about what temperature do the enzymes in photosynthesis denature completely?

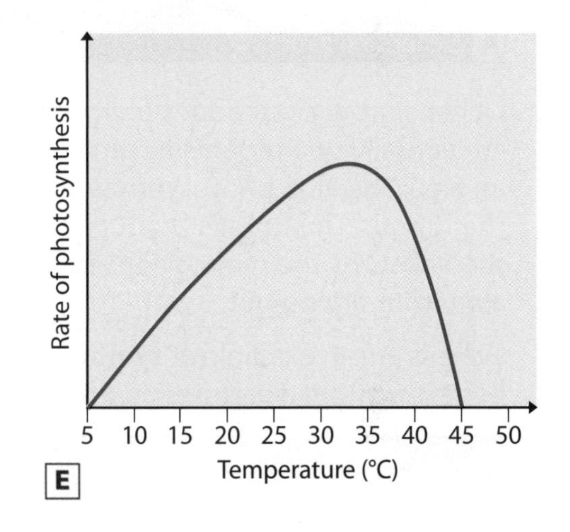

Greenhouses are used for growing some crops, so that the factors affecting the rate of photosynthesis can be controlled. Lights can be turned on and paraffin heaters are used to increase the temperature and produce carbon dioxide.

Like all fossil fuels, paraffin produces carbon dioxide when burnt. Most scientists agree that burning fossil fuels for heat and energy has dramatically increased the levels of carbon dioxide in the atmosphere. Scientists are trying to discover whether the increase in carbon dioxide has also caused an increase in photosynthesis.

> **?** **5** In an experiment, a 70% increase in carbon dioxide produced 40% more cotton from cotton plants.
> **a** Why was more cotton produced?
> **b** How might these plants have been grown?
> **c** What would a farmer consider when deciding whether to grow plants in this way?
>
> **6** Look at graph E.
> **a** What is the optimum growth temperature for this plant?
> **b** Different plants have enzymes with different optimum temperatures. Sketch a graph similar to graph E for a mountain plant.

Summary

You have been asked to plan a city community greenhouse, allowing people to grow crops like tomatoes. You need to present your ideas to the council. Write brief notes to summarise what you would say, and sketch diagrams to use in your presentation.

Symbiosis, commensalism, parasites

What are symbiosis, commensalism and parasitism?

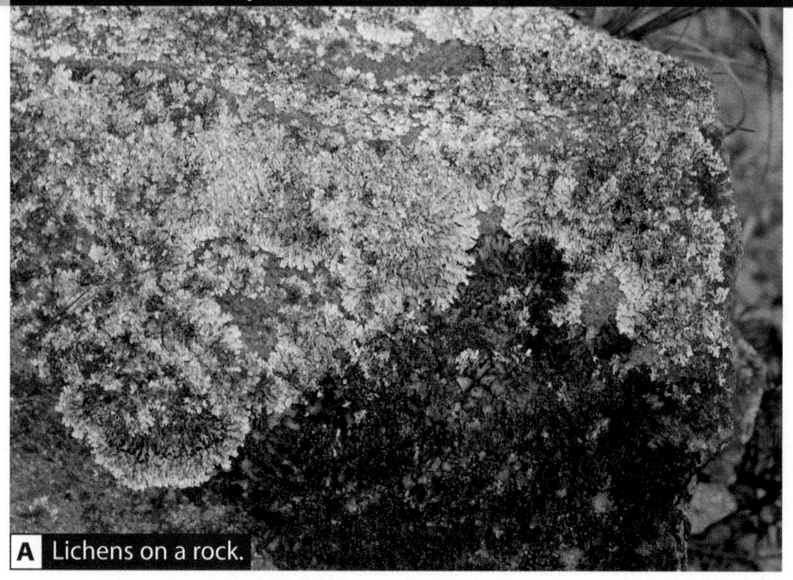

A Lichens on a rock.

Lichens grow as splodges on many buildings and trees. They are actually two organisms growing together – a fungus and an alga. The alga photosynthesises and provides glucose for the fungus. The fungus produces acid to attack the rock and then absorbs the mineral ions produced. It also protects the alga from drying out.

Lichens are an example of **symbiosis** – two organisms both benefiting from a partnership. Photo B shows another example.

B Clownfish are protected from predators by the sea anemones' stingers (the fish have a slimy covering that prevents them being stung). In return, the anemones are cleaned of parasites and get scraps of food left by the fish.

In a partnership called **commensalism**, one organism benefits and the other gets no benefit but is not harmed. An example is the imperial shrimp, which moves from one feeding ground to another by hitching rides on starfish.

?

1 Fungi normally feed on rotting matter. Why don't lichens need to be on rotting matter?

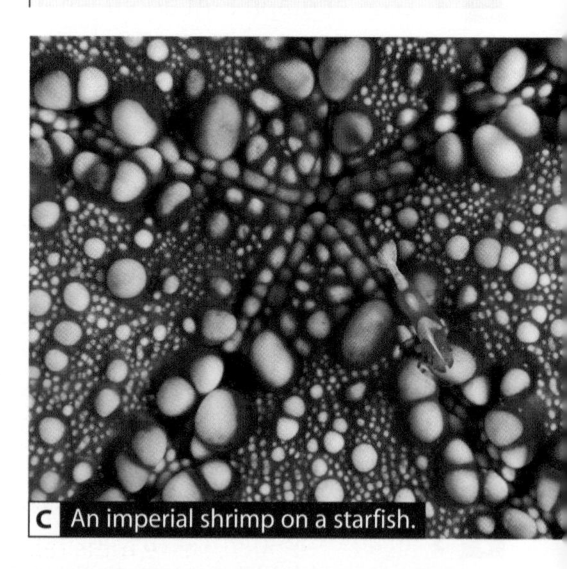

C An imperial shrimp on a starfish.

?

2 What is the difference between symbiosis and commensalism?

3 Red Sea goatfish dig holes in the sand to find invertebrates to eat. Splendour wrasse follow the goatfish, eating invertebrates that the goatfish miss. Is this symbiosis or commensalism? Explain your reasoning.

In **parasitism** one organism (the **parasite**) benefits but the other organism (the **host**) is harmed.

Tapeworms are flat worms that live in vertebrate intestines and can grow to 18 metres! A tapeworm's body is simply a head and sections containing reproductive organs that allow it to produce fertilised eggs. The eggs leave the host animal in its faeces and can then be swallowed by other animals, inside which they will hatch and grow.

Tapeworms cause harm by absorbing their hosts' nutrients, and this can result in malnutrition and severe weight loss. Tapeworms can also block up the intestines.

Hooks and suckers are used to attach it to the intestine wall. It has no use for eyes so it doesn't have any.

The flat body gives it a large surface area for food absorption. It has no need for a circulatory system, intestines or much of a nervous system.

The outside of the worm contains substances that stop digestive enzymes working – so that the worm is not digested.

Each section contains male and female reproductive organs so the worm can self fertilise its eggs if other tapeworms are not nearby.

D Tapeworms are so well adapted to their hosts that they don't waste energy making organs they don't need.

Mistletoe berries contain a sticky pulp that sticks to the beaks of birds that eat them. The birds scrape the pulp off their beaks, leaving the seeds stuck to new tree branches. Some seeds pass through birds' intestines and are deposited on branches in their droppings. A germinating mistletoe seed grows roots through the bark of a branch. These roots tap into the tree's veins to absorb water and nutrients, gradually weakening the tree.

?

6 What features of mistletoe berries allow the seeds to be spread easily?

7 From these two pages, name a symbiotic organism that eats parasites.

8 Dodder is a parasitic plant that has no leaves. Compared with mistletoe of a similar size, do you think it would do more or less damage to a host plant? Explain your reasoning.

?

4 a It is possible that in the early 1900s some 'dieting pills' contained tapeworm eggs. It is unclear whether this is true, but do you think they would work? Explain your reasons.

b Do you think this is a good way to diet? Explain your answer.

5 Why don't tapeworms need intestines or blood?

E Mistletoe clumps in a tree.

Summary

Write encyclopedia definitions for commensalism, parasitism and symbiosis, and give an example of each.

Parasite problems

What problems do parasites cause?

Between 1918 and 1919 a strain of flu killed around 50 million people worldwide. Flu is caused by viruses, all of which are parasites relying on the cells of other organisms to survive. Humans are hosts for many other parasites, including worms, bacteria and **protozoa** (another group of single-celled organisms).

Disease	Symptoms	Parasite	Deaths per year
Dengue fever	headaches, fever, severe joint and muscle pain, rash	*Flavivirus* viruses	48 000
Malaria	headaches, muscle aches, vomiting, high fever leaving person drenched in sweat	*Plasmodium* protozoa	1 300 000
River blindness	severe itchy skin, swollen legs and arms, spotty skin, eventual blindness	*Onchocerca* worms	0
Schistosomiasis	rash, itchy skin, fever, chills, cough, muscle aches	*Schistosoma* flatworms	15 000
Sleeping sickness	extreme tiredness, fever, severe headaches	*Trypanosoma* protozoa	48 000
Tuberculosis	bad cough, coughing up blood, tiredness, weight loss, sweating at night	*Mycobacterium* bacteria	1 566 000

A Some human diseases caused by parasites.

?

1 Name two human diseases caused by protozoan parasites.

2 How could a doctor tell the difference between people with dengue fever and malaria?

Some parasites damage food supplies. Nagana is a form of sleeping sickness that can kill cattle. Tapeworms cause cows to become weak and to produce less milk. Plant parasites cause crops to fail. Many diseases prevent people from looking after their crops, and food becomes scarce.

?

3 **a** Name a parasite that does not kill humans but can cause death indirectly.
b Explain how it can kill.

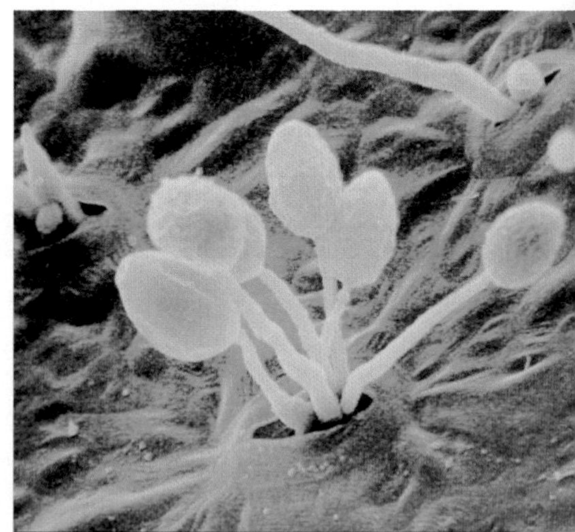

B A microorganism called *Phytophthera infestans* growing out through potato stomata. It kills potato plants and in the 1840s caused the starvation of nearly a million people in Ireland who depended on potatoes.

A host evolves over thousands of years, but so do its parasites. As the human immune system has evolved to defeat many parasites, *Schistosoma* flatworms have evolved to make use of it. Specific white blood cells coat the worms' eggs, causing them to be pushed through the intestine wall into the intestines. They then leave the body in the faeces. This process does not happen in people with weak immune systems!

?

4 How does the immune system help spread schistosomiasis?

H Parasites and natural selection

In 1954 A. C. Allison infected 30 volunteers with *Plasmodium*. 15 volunteers had one **allele** that forms abnormal **haemoglobin** (which carries oxygen in red blood cells). When this haemoglobin loses oxygen to body tissues, it causes the red blood cells to become sickle-shaped. The other 15 did not have the allele. Table C shows the results.

	Got sick	Did not get sick
People with one sickle-cell allele	2	13
People with no sickle-cell alleles	14	1

C The results of Allison's experiment.

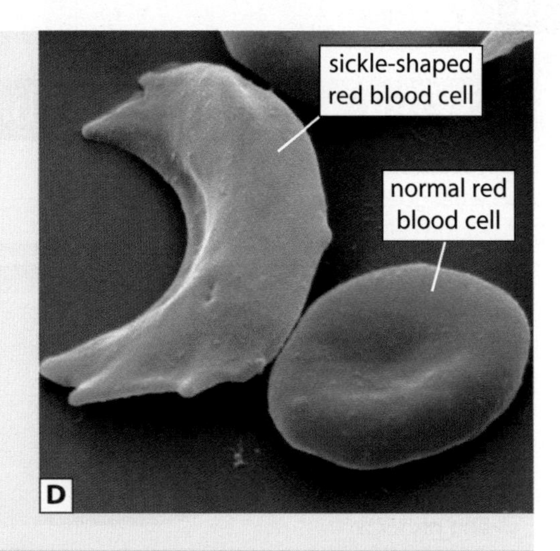

D

If someone inherits two sickle-cell alleles, they get **sickle-cell anaemia**. They suffer from tiredness, shortness of breath, swellings, and episodes of terrible pain when the sickle-shaped red blood cells stick together and block blood vessels.

In Africa, people with sickle-cell anaemia usually die before they can have children, so you might expect that the allele would gradually disappear from the population. However, children with one copy of the sickle-cell allele (**carriers** of the disease) are more likely to survive malaria than those without. These children are more likely to have their own children and pass on the allele. This is **natural selection** – a factor in the environment makes an allele more common than it would otherwise be.

?

5 **a** What disease did Allison's volunteers get sick with?
b What was the effect of having one sickle-cell allele?

6 What are the symptoms of sickle-cell anaemia?

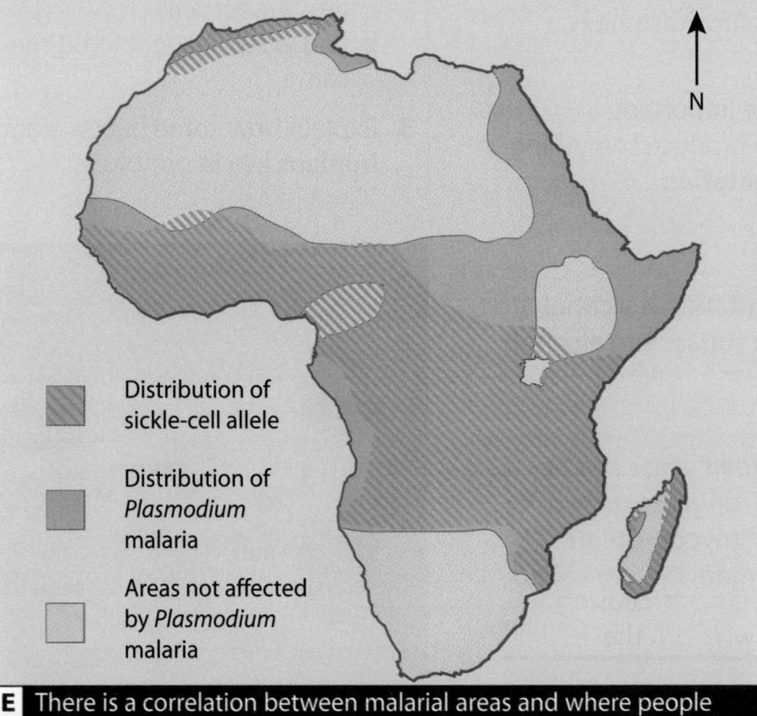

E There is a correlation between malarial areas and where people have the sickle-cell allele.

?

7 What is a sickle-cell carrier?

8 **a** What activities should people with sickle-cell anaemia avoid?
b Why should they avoid these activities?

9 In sickle-call anaemia, red blood cells are destroyed more quickly than normal red blood cells. Explain the symptoms this may cause.

Summary

Answer the question at the top of page 28 as a list of bullet points.

B7.11

Microorganisms and their uses

What are the uses of bacteria and fungal microorganisms?

cell membrane controls what goes into and out of the cell

circular **chromosome** made of DNA

soft **cell wall** holds the cell together and protects it

plasmids – small circles of DNA that also contain genes to make proteins

cytoplasm – where the cell's reactions occur

A Structure of *Yersinia pestis*. All bacteria have a similar structure.

During the 14th century, bubonic plague swept through Asia and Europe, killing over 200 million people. There were further big outbreaks until about 1700, and today there are still several thousand cases each year worldwide. It is caused by a bacterium called *Yersinia pestis*.

However, many bacteria are very useful. There are about 1000 different species living in an adult's body, with a total mass of about 2 kg! Many of these help us to digest food.

Some types of fungi, such as **yeast**, are also single-celled organisms. These cells are more complex than bacterial cells and have **chromosomes** in a nucleus. A chromosome is a large coiled molecule of DNA.

Bacteria and fungi can be grown to produce important products. When a microorganism is used to produce something on a large scale, the process is called **fermentation**.

Alcohol

One of the best known examples of fermentation is alcohol manufacture, where yeast are used to turn sugars into alcohol.

Single-cell protein

At the end of alcohol fermentation, the dead yeast can be used to make yeast extracts. Some other fungi are grown specifically to make food, such as Quorn® mycoprotein. Protein-rich foods obtained from microorganisms are called **single-cell proteins**.

?

4 Write dictionary definitions for these words: fermentation, single-cell protein, yeast.

?

1 Draw a table of bacterial cell features and their functions.

2 Name a feature that:

a is different between bacterial and fungal cells

b you would expect to be the same.

3 Explain how some bacteria and humans live in symbiosis.

B All of these were made by microorganism fermentation.

Enzymes

Enzymes are used to produce cleaning products, textiles, paper, some plastics and many foods. For example, **rennin** is an enzyme used in cheese making and is responsible for turning milk into a mixture of solid lumps and liquid. The solid lumps are compressed to make cheese. Rennin is traditionally obtained from calf stomachs but can also be extracted from a type of fungus and from genetically modified bacteria (see page 32). Rennin from these two sources is used in 'vegetarian cheeses'.

Antibiotics

Both bacteria and fungi can produce chemicals that stop other bacteria growing. These are **antibiotics**. Diagram D shows how antibiotics are produced using fermentation.

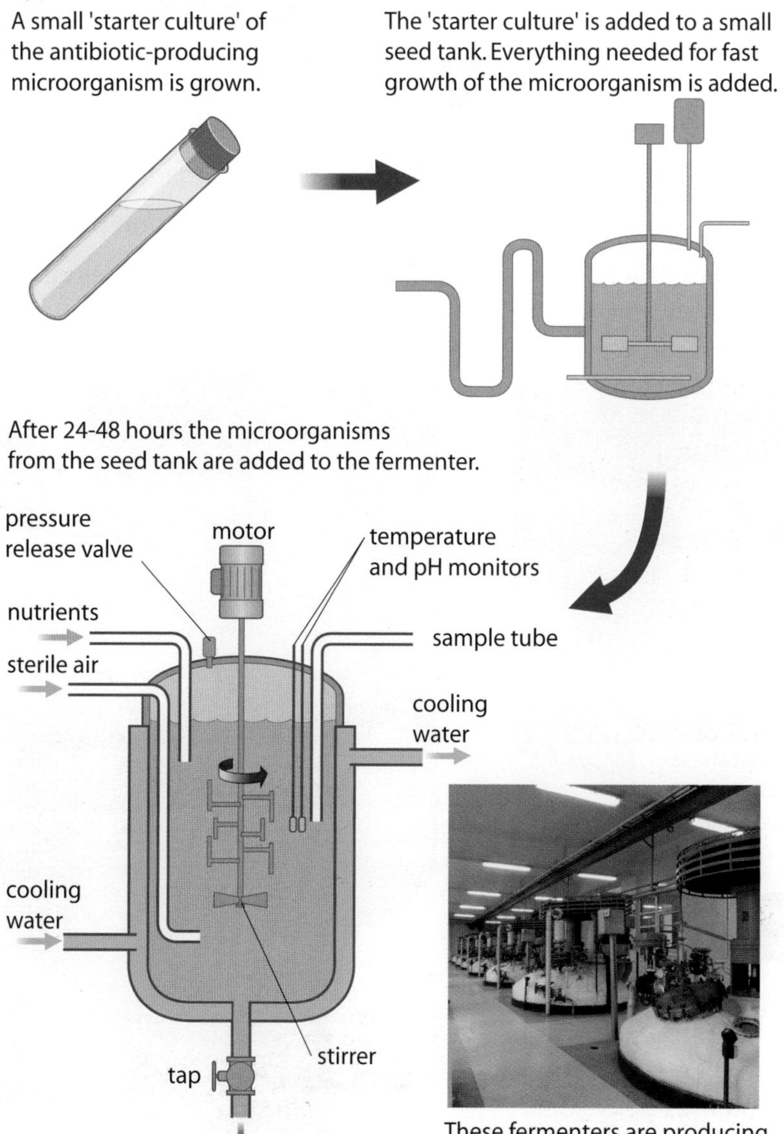

These fermenters are producing antibiotics for farm animals.

C *Penicillium* mould fungus produces an antibiotic (penicillin) that kills bacteria.

?

5 Look at photo C. Why is there a clear ring around the *Penicillium*?

6 Why are antibiotics not used to treat dengue fever? (*Hint*: look back at page 28.)

7 **a** Give an example of an enzyme made using fermentation.
b Describe briefly how this enzyme would be produced by fermentation.

8 Name one possible nutrient that needs to be supplied to the microorganisms used to make antibiotics.

9 In diagram D, sterile air is supplied to the fermenter tank.
a Why does air need to be supplied?
b Why must it be sterile?

10 Explain why a jacket of cooling water surrounds the fermenter.

Summary

You have been asked to produce a leaflet entitled 'Useful microorganisms' to be given out at a food fair. Write a list of things you would include.

Genetic modification

How are organisms genetically modified?

About 90% of cheese is made using rennin from genetically modified (GM) bacteria. Rennin is made in the stomachs of young farm animals, like calves. The **gene** for making rennin was isolated and put into bacteria, which now make rennin.

A Cheese made using rennin from GM bacteria.

Isolation

To make a GM organism, scientists must find the gene they want by a process called **isolation**. One way is to make a **gene probe**. This is a short single strand of artificial DNA that contains some of the DNA code needed to produce a useful protein. The probe is added to the organism's DNA where it hopefully sticks to a matching section of DNA, showing scientists where the whole gene is. The gene for the useful protein can then be cut out of the organism's DNA.

?

1. What do you think are the benefits of using rennin from GM bacteria?
2. What is a GM organism?
3. What is 'gene isolation'?
4. In replication, what is being replicated?

Replication

Many copies of a gene are needed for **genetic modification**, and they are made by a process called replication. Often the isolated gene is put into the DNA of a bacterium. The bacteria grow and multiply, their DNA is then extracted and the genes are cut out. PCR machines can also make copies of genes.

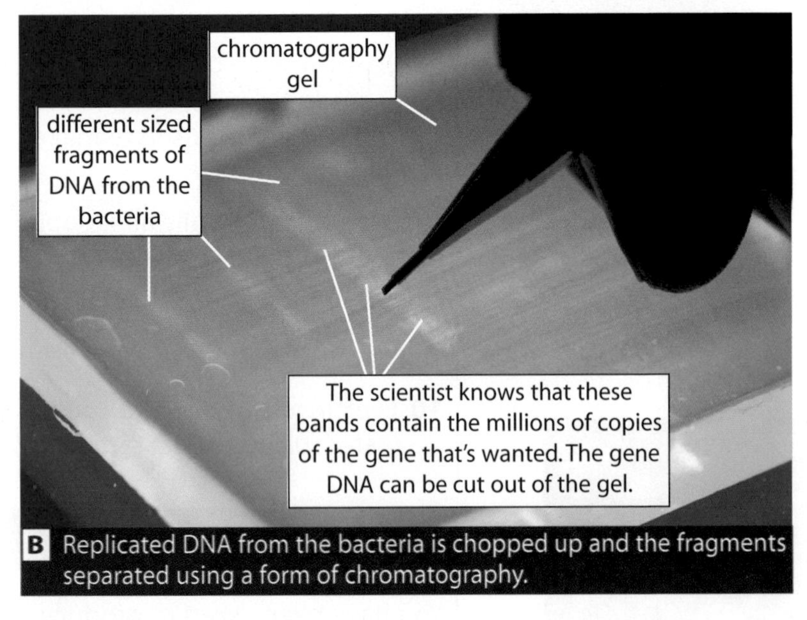

B Replicated DNA from the bacteria is chopped up and the fragments separated using a form of chromatography.

Gene transfer

One way of getting the replicated gene into a new organism (**gene transfer**) is to use a gun! A gene gun fires gene-coated pellets at cells. If the gene gets into a cell, it can be added to the cell's DNA.

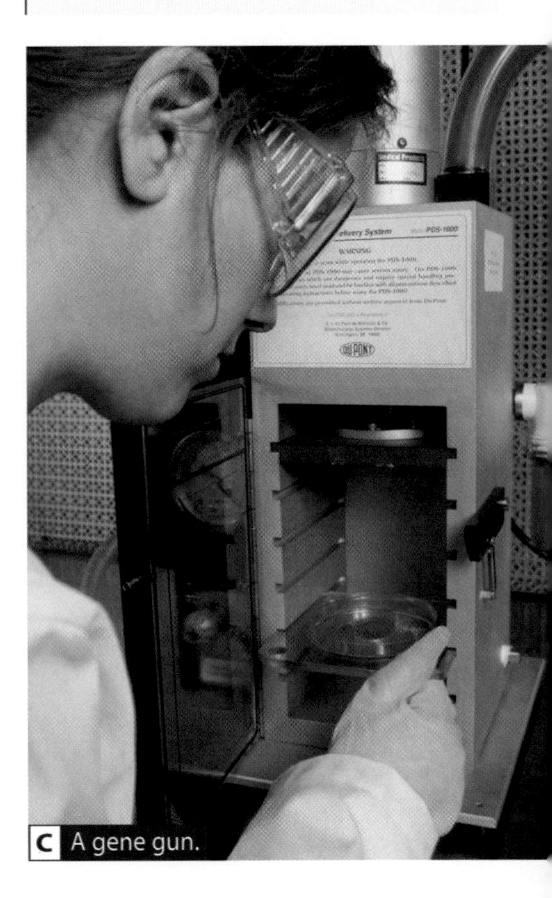
C A gene gun.

H Using vectors

Plasmids are small circles of DNA found in the cytoplasm of bacteria, containing genes for a small number of proteins. Plasmids can be cut open and new genes inserted into them. The new plasmids are added to a liquid containing the bacteria, and an electric shock causes the bacteria to take in the plasmids. Useful genes can also be put into viruses, which then attack cells and add the useful genes to the DNA of the cells. Something that carries new DNA into a cell is called a **vector**.

?

5 Explain how you think the bacteria used to make rennin were modified.

6 Viruses that are used in gene transfer are often genetically modified themselves to *not* do something. What should they be prevented from doing?

Examples of genetic modification

GM bacteria produce many useful proteins, like enzymes, medicinal drugs and hormones (e.g. insulin, needed by diabetics). GM plants can be more resistant to diseases and produce greater yields. GM animals can be faster growing, have better meat or produce useful proteins in their milk.

This GM cotton is resistant to a pest called bollworm.

D

Golden rice has been genetically modified to provide people with more vitamin A often missing from the diets of poor rice farmers.

E

GM Jersey cows can produce human growth hormone in their milk – to treat people with a deficiency of the hormone.

F

The GM debate

Some people think that GM crops and animals may help reduce hunger in developing countries. However, others think that GM organisms will reproduce with wild organisms, producing organisms that could get out of control.

?

7 Name two ways in which each of the following organisms could be genetically modified to be more useful:
- **a** cows
- **b** tomato plants.

8 In the UK foods containing GM organisms or parts of GM organisms have to be labelled to say that they are 'GM'. The cheese in photo A is not labelled 'GM'.
- **a** Explain why not.
- **b** Do you think it should be labelled? Explain your reasons.

Summary

Draw and label a flowchart to summarise how GM organisms are made.

B7.13 Genetic testing

What is genetic testing?

Similar techniques to those used in genetic modification are used in **genetic testing** – testing people for alleles that might cause disease. For instance, people might want to know whether they are carriers for sickle-cell disease (see page 29).

A These newlyweds had genetic tests to make sure that they were not carriers for Tay–Sachs disease, because they didn't want to risk having children with the disease.

? **1** Suggest why someone might want to know whether they are a carrier for sickle-cell disease.

Isolating the DNA

A DNA sample for a genetic test is obtained from blood. Chemicals split open the blood cells and their nuclei, releasing the DNA. (Red blood cells contain no nuclei, so all the DNA comes from white blood cells.) The DNA is cut into small fragments by enzymes, which act like chemical scissors.

? **2** Why does DNA from a blood sample come only from white blood cells?

The gene probe

Scientists know the order of the bases (G, C, A and T) in most human genes and also how this order differs in some alleles that cause disease.

part of the sickle-cell allele

CTG ACT CCT GTG GAG AAG TCT

GAC TGA GGA CAC CTC TTC AGA

part of allele for normal haemoglobin

CTG ACT CCT GAG GAG AAG TCT

GAC TGA GGA CTC CTC TTC AGA

B Parts of the sickle-cell allele and the normal haemoglobin allele.

? **3** Look at diagram B. How does the sickle-cell allele differ from the normal haemoglobin allele?

A gene probe is a short *single* strand of DNA with a marker attached. It is made by a machine and contains a small part of the base sequence from an allele that causes a disease. Diagram C shows how the gene probe is used. The probe sticks to a single DNA strand that has complementary bases, and it is found by looking for the marker attached to it.

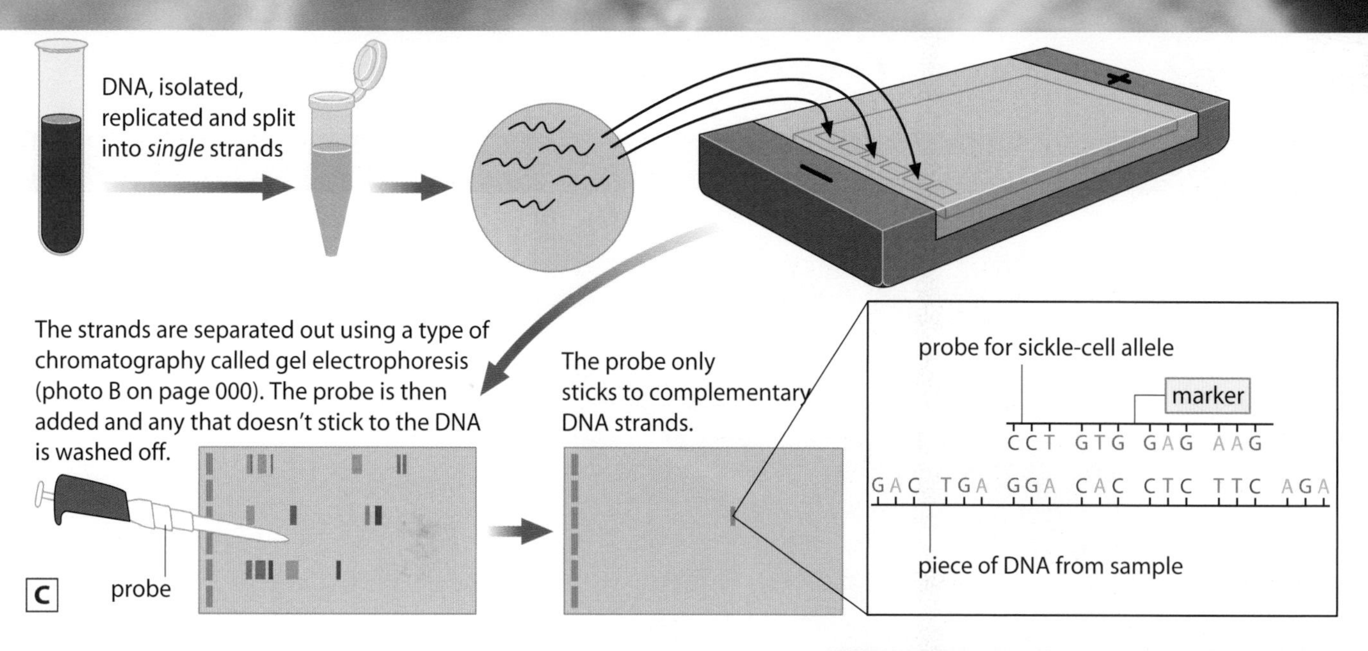

C

Gene probe markers

There are two common ways of marking a probe, to see where it goes. The first is to attach a chemical that glows under ultraviolet light. The other way is to make the probe using radioactive phosphorus atoms (which form part of the DNA backbone). If you place photographic paper or X-ray film on some separated pieces of DNA, the radiation creates a dark spot showing which piece of DNA the probe has stuck to. This is called **autoradiography**.

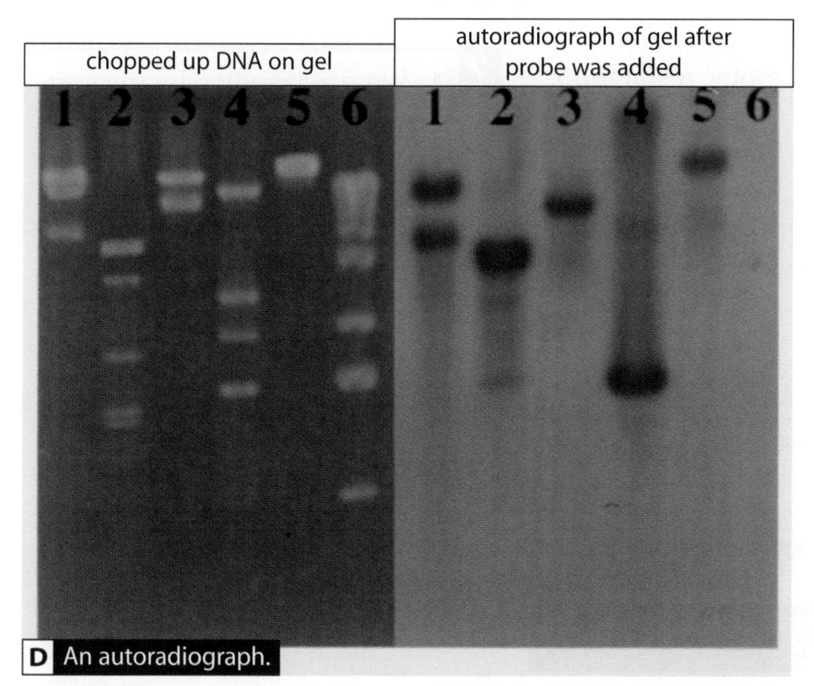

D An autoradiograph.

DNA chips

By putting different probes on a piece of silicon, it is now possible to test for many alleles at once. In this process, the sample DNA has the markers attached.

?

4 Write out the bases in a probe to find the allele for normal haemoglobin.

5 **a** In the autoradiograph in photo D, which of the six DNA samples does not contain the allele on the probe?

b What would the autoradiograph look like if the probes that did not stick were not washed off?

6 Probes are usually longer than the one shown in diagram C. Why do you think they need to be longer than this?

7 Find out what Tay–Sachs disease is.

E DNA probes on a piece of silicon – a so-called DNA chip.

Summary

Write step-by-step instructions on testing for Tay–Sachs disease.

B7.14 Energy and aerobic respiration

What is aerobic respiration?

A Energy from respiration is needed so the cells in this person's muscles can **contract**.

All cells need energy for growth (building of new cells), chemical reactions and movement. The energy for these processes comes from respiration. Respiration is a series of reactions that result in releasing energy from glucose. They can be summarised in this word equation:

glucose + oxygen → carbon dioxide + water (+ energy released)

This is aerobic respiration because it uses oxygen that comes from the air. Most cells carry out aerobic respiration.

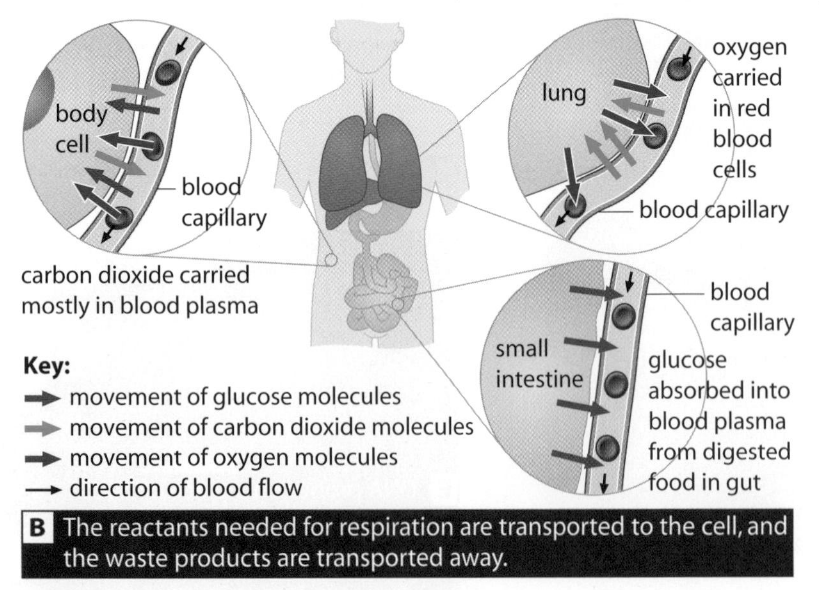

B The reactants needed for respiration are transported to the cell, and the waste products are transported away.

The circulatory system is important for respiration. The speed at which the blood flows around the body is affected by the heart beat rate. The faster the blood flows, the quicker materials can be exchanged between the blood and cells. Materials are also exchanged faster if **blood pressure** is higher.

?

1. Write a definition for the term aerobic respiration.
2. Explain why cells carry out aerobic respiration.

?

3. Describe where the materials for respiration enter the body, and how they reach all cells.

Heart rate and blood pressure vary naturally between individuals, because they depend partly on body size, gender and fitness. Your heart rate and blood pressure may also vary with different levels of activity or if you are anxious. So 'normal' rates and pressures are given as ranges.

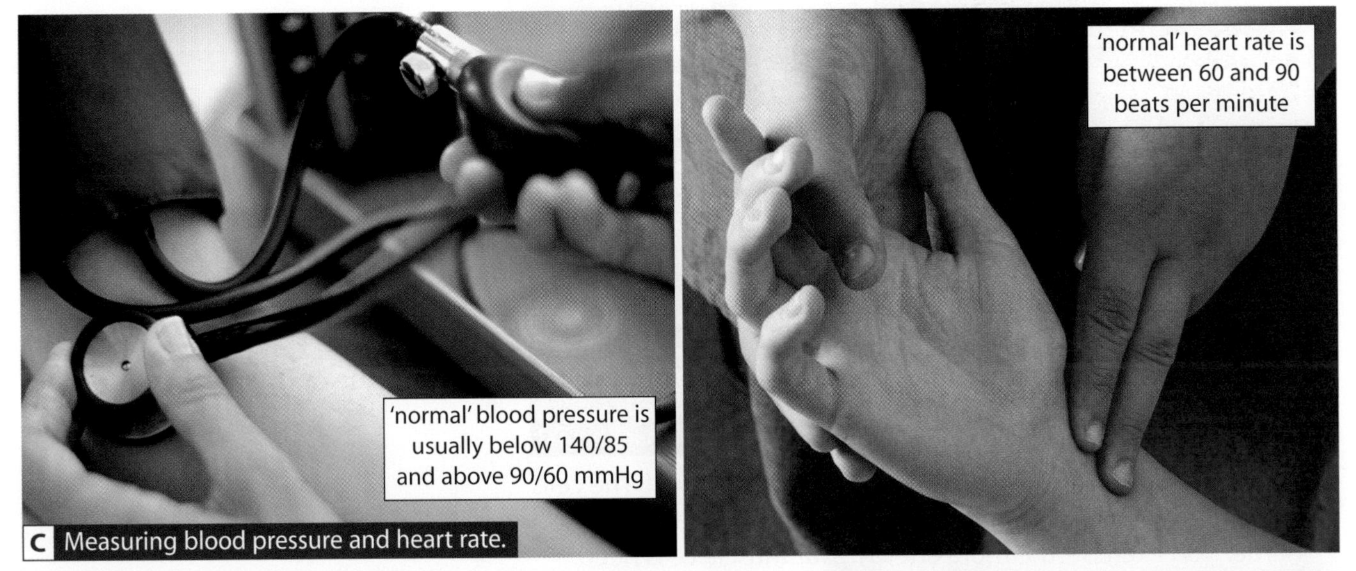

C Measuring blood pressure and heart rate.

H During respiration, the breakdown of each molecule of glucose releases energy. The energy is used to build many small molecules of a substance called **adenosine triphosphate (ATP)**.

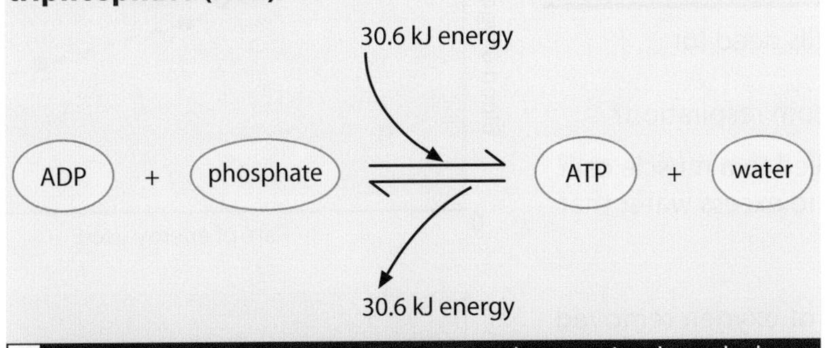

D Energy is used to convert ADP to ATP, and energy is released when ATP is converted back to ADP.

The advantage of transferring the energy to ATP is that it can be released again in much smaller amounts. The energy from the breakdown of ATP is used to power many different reactions in cells. This is why ATP is often referred to as the 'energy currency' of living things. Every day you convert over half your body weight of ATP from or back to ADP.

?

6 Give as many advantages as you can of using the energy released from the respiration of different food chemicals to make ATP.

7 Suggest why ATP can be referred to as the 'energy currency' of living things.

?

4 Explain why 'normal' blood pressure and heart rate are given as ranges.

5 Craig's doctor measured his heart rate. It was 95 beats per minute. Explain why the doctor told Craig to sit quietly for 10 minutes before measuring his heart rate again.

Summary

Look at photo A. Draw a labelled diagram to show what is happening in this person's body so that the muscle can contract. Include the words 'aerobic respiration' and 'circulatory system' in your labels to explain their importance.

H Explain the role of ATP in respiration in your labels.

B7.15 Exercise and aerobic respiration

How does the body change during aerobic exercise?

A How can these people get the energy they need to run for many hours?

When you exercise, you use more energy in movement. The cells in your body, particularly the cells in muscles, respire faster. The harder you exercise, the more energy your muscle cells need.

As cells respire faster, they need oxygen and glucose to be supplied faster. They also need to get rid of carbon dioxide and excess water faster.

?

1. **a** What are the two chemicals that cells need for respiration?
   b What are the two waste products from respiration?
2. **a** How are glucose and oxygen supplied to a muscle cell?
   b What happens to carbon dioxide and excess water that leave a cell?

When you begin exercising, the amount of oxygen removed from the blood by muscle cells increases. More glucose is also removed from the blood. However, if there isn't enough glucose in the blood for the increased level of respiration, chemicals stored in muscle and liver cells are converted to glucose to supply what is needed.

?

3. Explain why liver and muscle cells store chemicals that can be converted to glucose.
4. Describe what effect increasing the level of activity has on:
   a oxygen concentration in blood near cells
   b carbon dioxide concentration in blood near cells.
5. The graphs in diagram B show concentrations in the blood in veins. Explain why these measurements were taken in the veins, not the arteries.

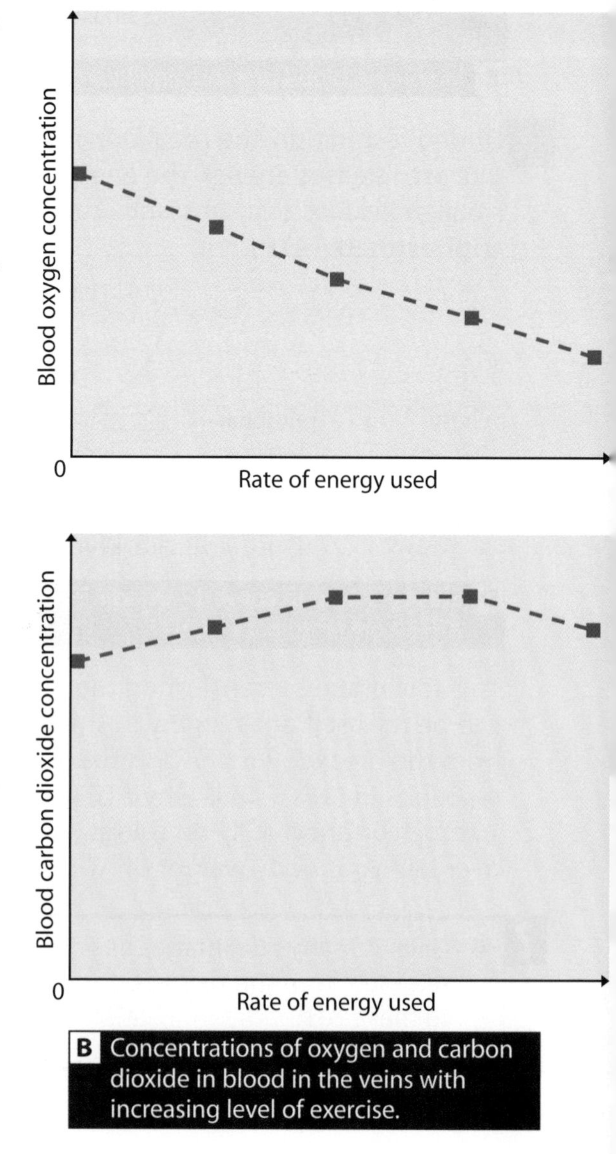

B Concentrations of oxygen and carbon dioxide in blood in the veins with increasing level of exercise.

When you exercise, your blood needs to circulate faster to supply all the extra oxygen that is needed, and to remove all the carbon dioxide and excess water from cells. This means your heart beats faster. The oxygen and carbon dioxide also need to be exchanged faster in the lungs, so you breathe faster and more deeply.

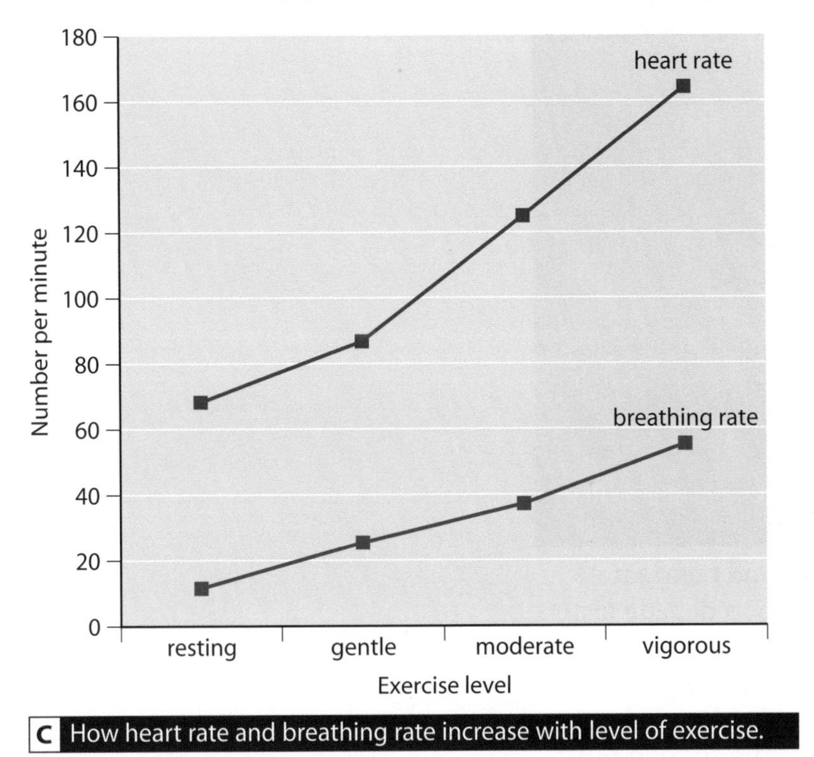

C How heart rate and breathing rate increase with level of exercise.

Exercise that increases heart rate and breathing rate is called **aerobic exercise**. Our bodies evolved in response to regular and frequent activity. Today, our lives don't require us to do very much exercise. In the UK, 7 out of 10 adults don't get enough exercise. This is leading to an increase in health problems, such as high blood pressure and heart disease.

?

6 Explain why, until about the last 50 years, physical activity was a large part of people's lives.

7 Explain why regular aerobic activity is important for keeping healthy.

8 The rate of diffusion across a membrane, such as between a blood capillary and the inside of the lung, is affected by the difference in concentration on each side of the membrane. Explain how breathing faster and deeper increases the rate of exchange of oxygen and carbon dioxide between the blood and air in the lungs.

9 Heart rate in someone who is very fit increases less rapidly with exercise than in someone who is less fit. Suggest reasons for this.

D Many people now go walking to get more aerobic exercise.

Summary

Draw a concept map to help you answer the question at the start of this topic.

Anaerobic respiration

When do we use anaerobic respiration?

A This athlete is exercising so hard that she needs more energy than she can get from aerobic respiration.

As you exercise faster, your breathing rate and heart rate increase to supply oxygen and remove carbon dioxide faster. But there comes a point when they can't increase further. If you exercise harder than this, your body needs to get the extra energy it needs in another way. **Anaerobic respiration** provides this extra energy. It also provides the sudden burst of energy that you need when you start exercising hard.

Anaerobic respiration also breaks down glucose, but it doesn't need oxygen and it doesn't make the same products. The equation for anaerobic respiration is:

glucose → lactic acid (+ energy released)

The **lactic acid** released in anaerobic respiration builds up as exercise continues. Anaerobic respiration plays an essential part in keeping muscles working when there isn't enough oxygen getting to muscle cells.

?

1. Explain why anaerobic respiration has this name.
2. Draw up a table to show the similarities and differences between aerobic and anaerobic respiration.
3. Anaerobic respiration occurs mostly in muscle cells. Explain why.
4. If you start vigorous exercise suddenly, your muscles will need to get some of their energy from anaerobic respiration.
   a Explain why this happens.
   b Explain the importance of warm-up exercises.

B Warm-up exercises increase the heart and breathing rates so that more vigorous exercise can be done aerobically.

If you have been exercising anerobically, lactic acid will be left in your cells after your exercise rate has slowed down. The body uses oxygen to break down this lactic acid. You continue to breathe quickly and deeply for a while after exercise, to supply this extra oxygen. This is often described as 'paying back the **oxygen debt**'.

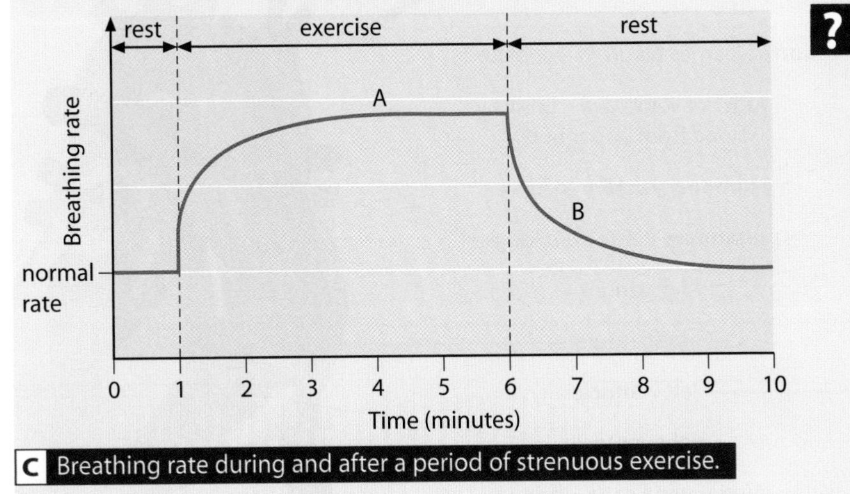

C Breathing rate during and after a period of strenuous exercise.

?

5 Look at graph C. Explain the breathing rate at:
   a point A
   b point B.

6 **a** Define oxygen debt.
   b Explain what is meant by 'paying back the oxygen debt'.

Anaerobic respiration is the partial breakdown of glucose. It releases much less energy from each glucose molecule than aerobic respiration. Its advantage is that it can continue to release energy from glucose when there is not enough oxygen for all the energy to come from aerobic respiration.

D Seals use anaerobic respiration during deep dives.

?

7 Draw up a table to show the advantages and disadvantages of aerobic and anaerobic respiration.

8 Anaerobic respiration can be used in most animals only as a supplement to aerobic respiration, not a replacement. Explain why.

9 Suggest how anaerobic respiration could help the following animals to survive. Explain your answers.
   a seal while diving
   b cheetah sprinting after prey

Summary

Your friend is planning to start running to get fit. Write her an e-mail explaining why she needs to do warm-up exercises first, and how running at different speeds will affect the kind of respiration her muscle cells will use.

B7.17 The circulatory system

How is blood pumped around your body?

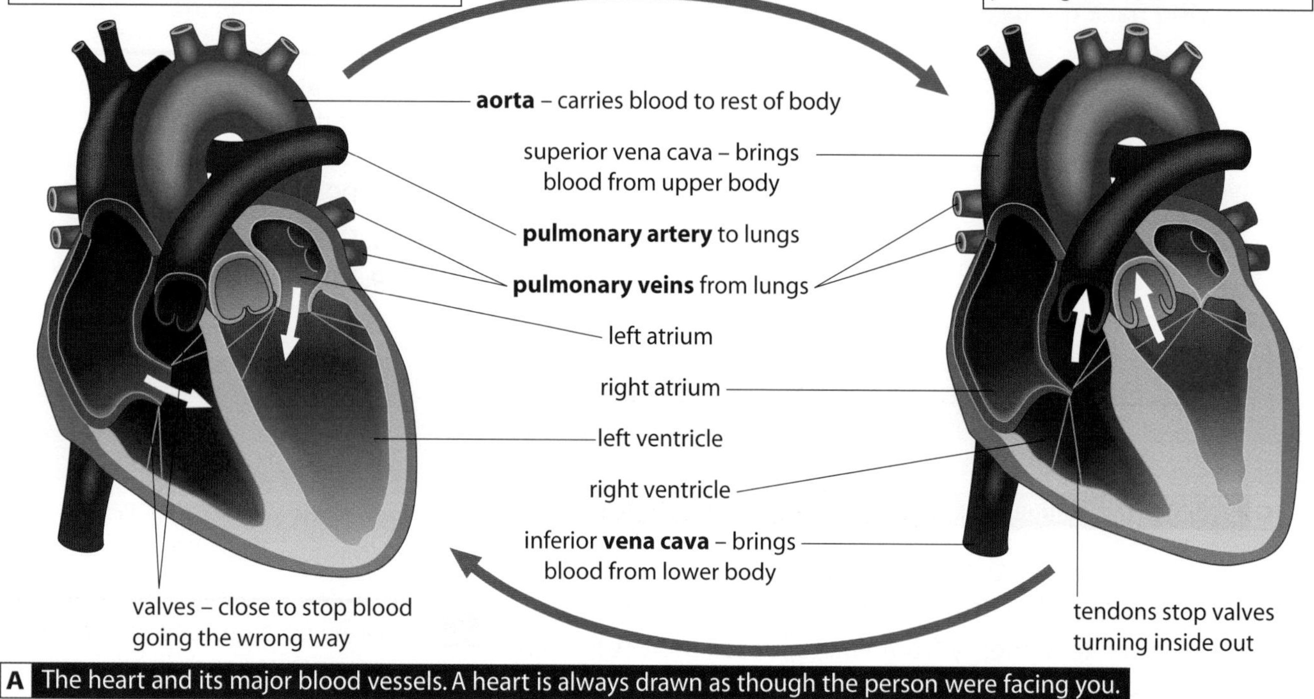

A The heart and its major blood vessels. A heart is always drawn as though the person were facing you.

Your blood travels around your body about once a minute, and that flow is kept going by your **heart**. The heart is divided into four **chambers** with **blood vessels** leading to and from it. The heart and blood vessels together are called the **circulatory system**.

A heartbeat

Your heart beats about 70 times per minute (your heart rate). During a heartbeat, the **atria** contract, squeezing the blood into the **ventricles**. These then contract, pushing blood out of the heart, and the atria refill with blood. **Valves** stop blood flowing the wrong way, and it is the sound of these valves shutting that you hear as 'lub-dub' when listening to a heart.

Blood vessels

There are nearly 100 000 km of blood vessels in your body, divided into three types. Blood flowing away from the heart is carried in thick-walled tubes called **arteries**. These feed into tiny **capillaries**, which have walls that are only one cell thick. Blood is taken back to the heart in **veins**, which contain valves to stop blood flowing the wrong way and help blood move up the body towards the heart, against gravity.

?

1. What are the names of the four heart chambers?
2. Which blood vessels carry blood away from the heart: veins or arteries?
3. When listening to a heart, the 'lub' sound is made by the valves between the heart chambers shutting. What makes the 'dub' sound?

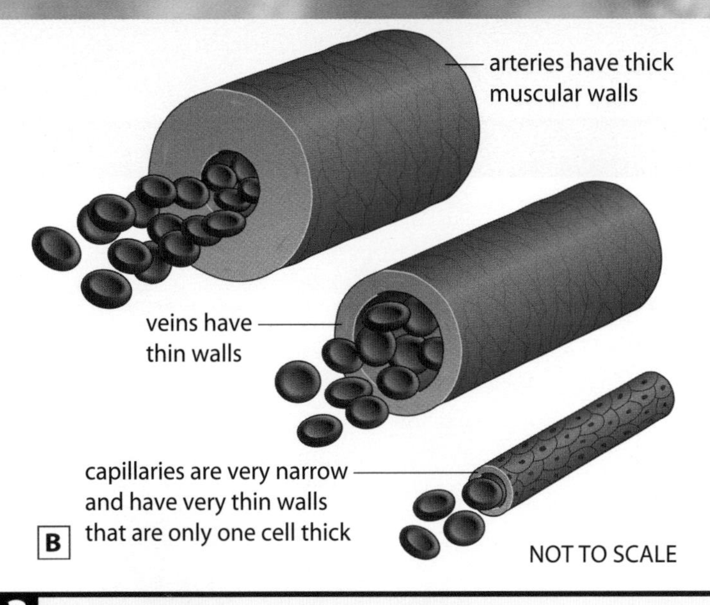

B

?

4 People with chronic venous insufficiency disease have vein valves in their legs that don't work properly. Suggest one symptom of this disease.

Pulse

When your heart pushes out blood, your aorta stretches. This causes a wave of stretching to travel down through the walls of the arteries. This is what you can feel as a **pulse**.

A

- How would you measure someone's pulse and their heart rates? Would you expect them to be the same? Why?
- Are male and female resting pulse rates about the same? How would you find out?

A double pump

The human heart has two sides. Each acts as a pump, pushing blood through a distinct set of blood vessels – one to the lungs and one to the rest of the body. This is called a **double circulatory system**. In the lungs the blood picks up oxygen, which causes it to change colour from dark red to bright red.

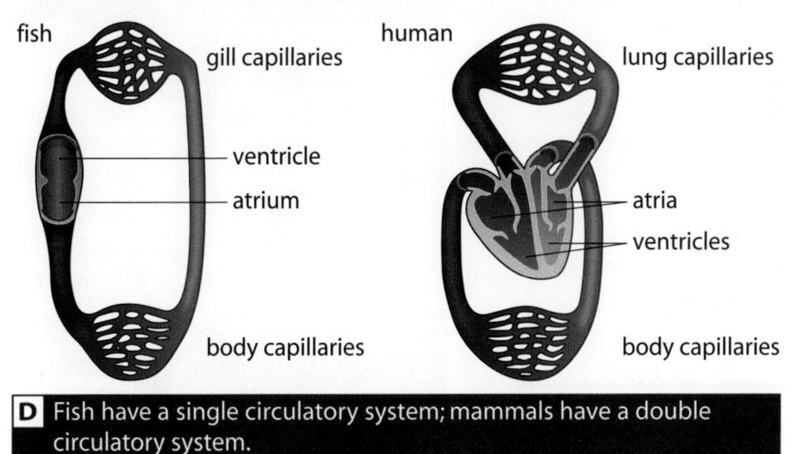

D Fish have a single circulatory system; mammals have a double circulatory system.

C Valves in veins.

?

5 A website says: 'You should feel a pulse … under your fingers; that is the movement of your blood.' What is wrong with this statement?

6 Why is the blood in veins a different colour to the blood in arteries?

7 Describe the route blood takes on its journey from a vena cava to the aorta.

8 Photo E shows someone with varicose veins.

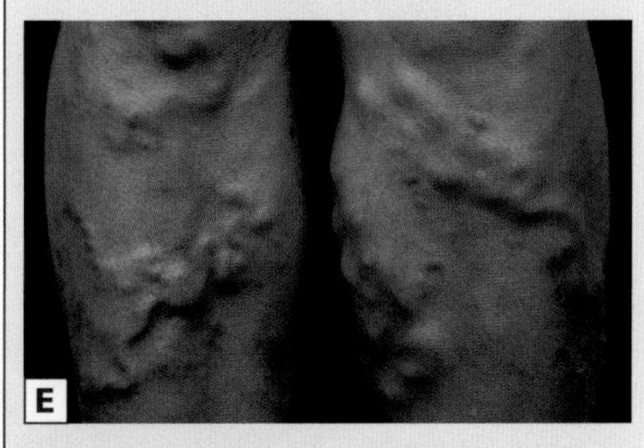

E

a Suggest how varicose veins might be caused.

b Find out if you are correct, using books or the internet.

Summary

Using the words in bold in this topic, draw a concept map to summarise what you know about the human circulatory system.

Blood

What are the functions of blood?

An average adult body contains about five litres of blood. In each *cubic millimetre* (mm^3) there are about 5 000 000 **red blood cells**, 7000 **white blood cells** and 250 000 **platelets**, all suspended in a straw-coloured liquid called **plasma**.

Red blood cells

Red blood cells have no nuclei and are packed with **haemoglobin**, a protein that combines with oxygen molecules to form **oxyhaemoglobin**. In the lungs, oxygen diffuses into the capillaries and attaches to the haemoglobin in red blood cells.

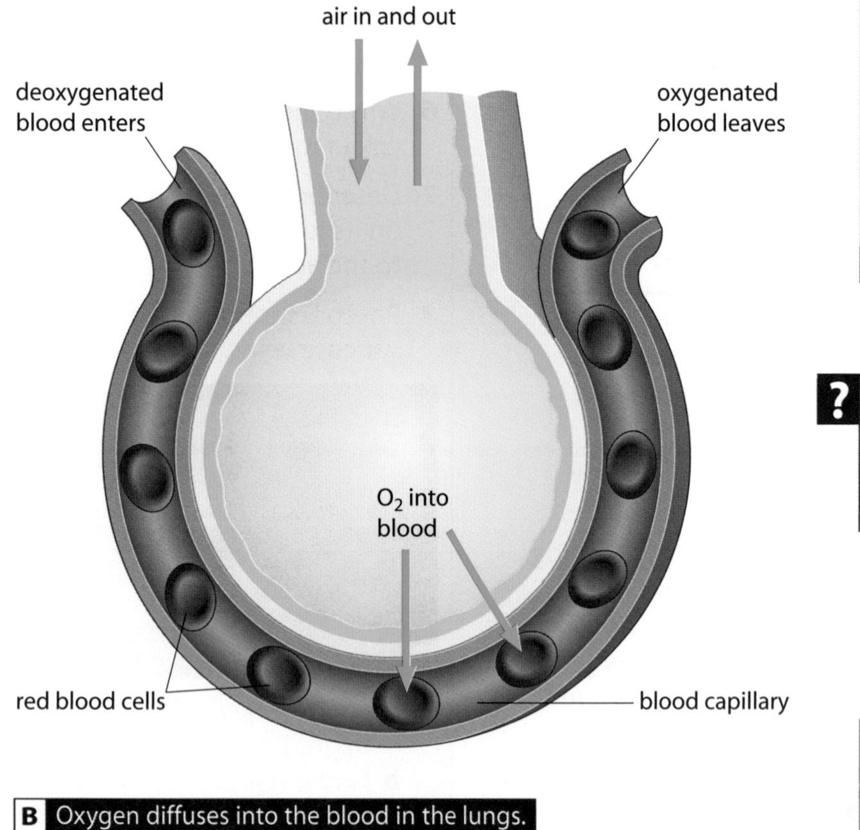

B Oxygen diffuses into the blood in the lungs.

White blood cells

White blood cells kill microorganisms that cause infections. Some produce **antibodies** that stick to microorganisms and help to kill them. Others are able to surround microorganisms and digest them.

?

3 Write a caption for photo C.

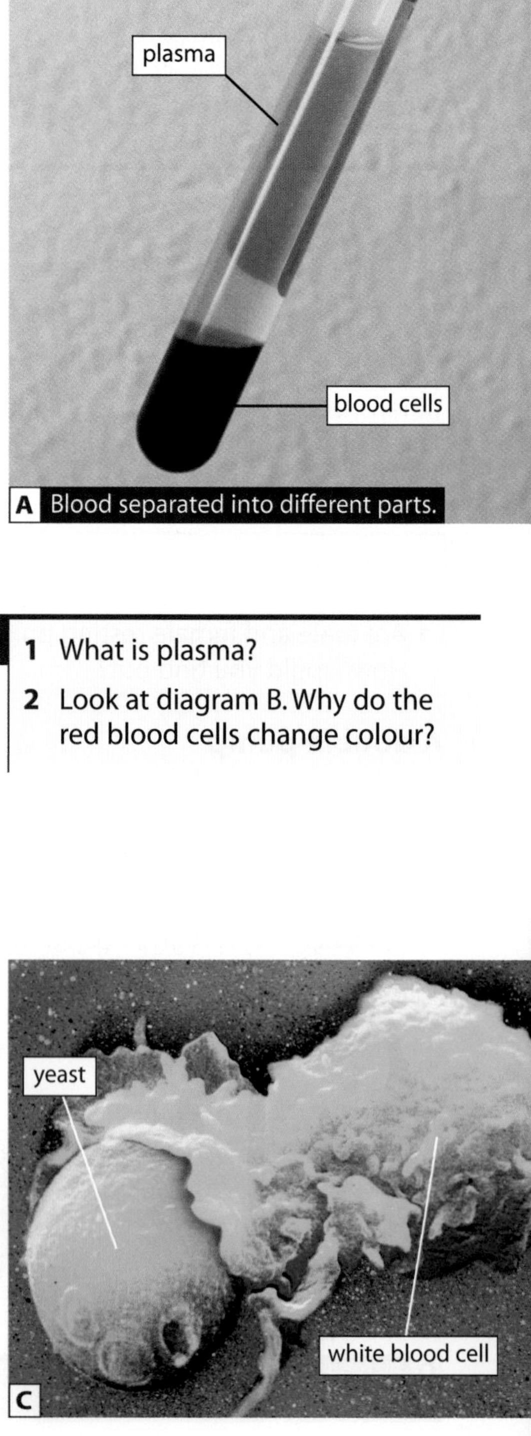

A Blood separated into different parts.

?

1 What is plasma?

2 Look at diagram B. Why do the red blood cells change colour?

C

Platelets

Platelets are tiny fragments of cells. They have no nuclei and produce chemicals needed to clot the blood at the site of an injury, for example when the skin is cut.

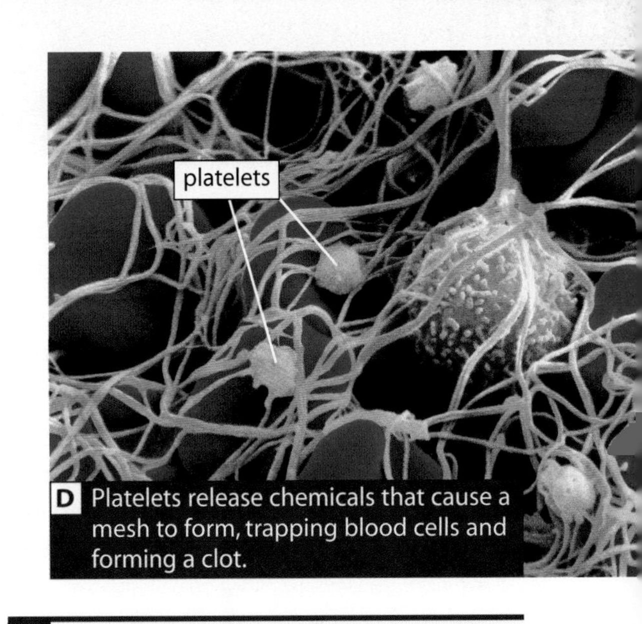

D Platelets release chemicals that cause a mesh to form, trapping blood cells and forming a clot.

? **4** What cells have been trapped to form the clot in photo D?

Tissue fluid

Chemical reactions in the cells use up glucose and oxygen, and produce waste such as carbon dioxide and **urea**. The **tissue fluid**, which surrounds cells, therefore has less oxygen and glucose and more carbon dioxide and urea than the blood in the capillaries. These substances diffuse into or out of the capillaries through the capillary walls.

In **capillary beds** (networks of capillaries in tissues) the high pressure in the arteries pushes plasma out through small holes in the capillary walls, adding to the tissue fluid. This helps to carry more glucose and oxygen from the blood to the cells.

Urea and carbon dioxide diffuse into the capillaries and are mostly carried in the blood dissolved in the plasma. The blood in the capillaries at the vein end of a capillary bed is under low pressure, so water from the tissue fluid moves by osmosis back into the capillaries.

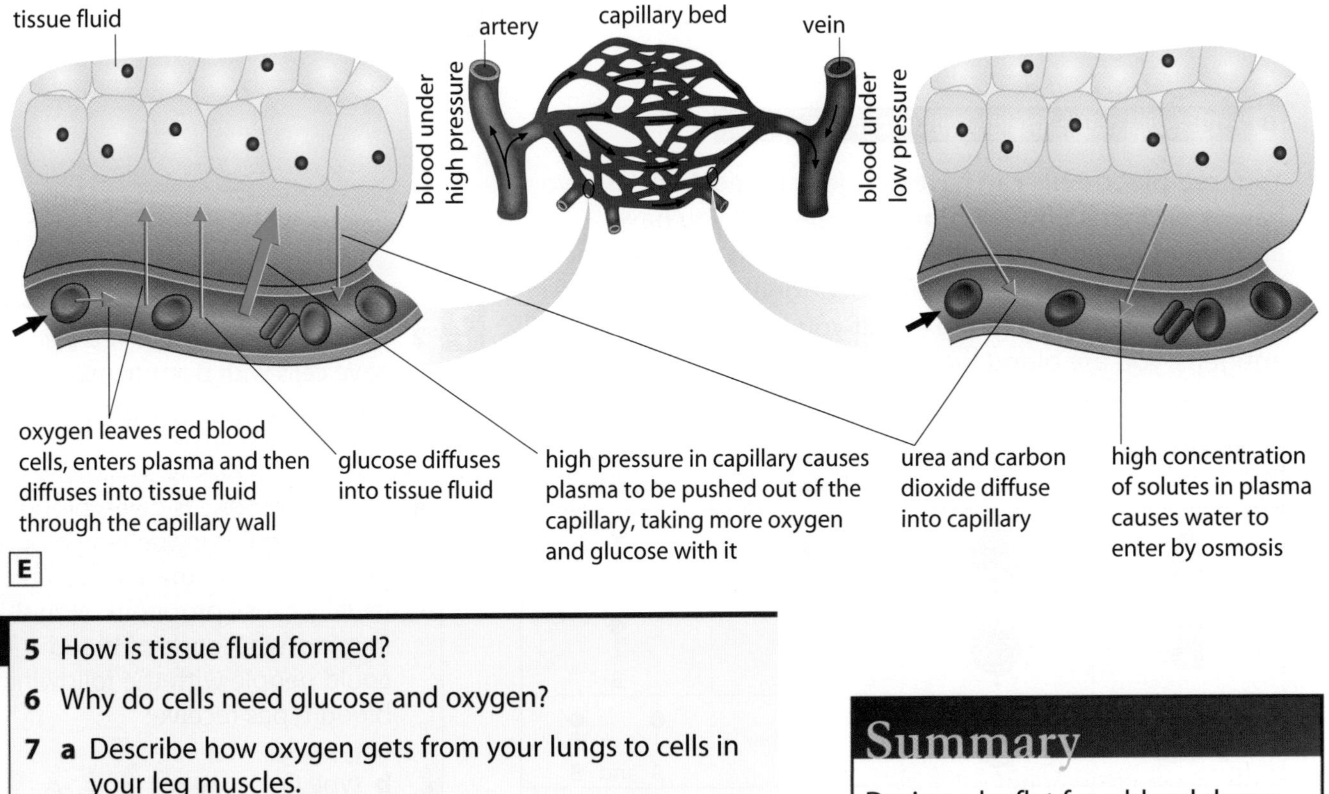

E

?

5 How is tissue fluid formed?

6 Why do cells need glucose and oxygen?

7 **a** Describe how oxygen gets from your lungs to cells in your leg muscles.

b Describe how carbon dioxide gets from your leg muscles to your lungs.

8 Suggest a reason why red blood cells have no nuclei.

Summary

Design a leaflet for a blood donor centre to show what each part of the blood does.

B7.19 Blood antigens

Why do people have different blood types?

Every day 4500 litres of blood are used in hospitals in the UK. However, you can't give everyone the same sort of blood – there are different types.

All of your cells have proteins called **antigens** sticking out of their surfaces. Your white blood cells recognise the antigens on your own cells but destroy cells with unfamiliar antigens.

One way that white blood cells destroy foreign cells is by making antibodies. These are found in the plasma and stick to foreign cells, clumping them together and helping to kill them.

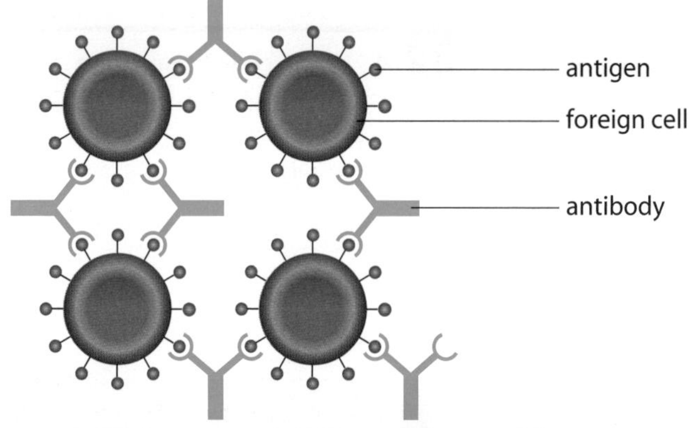

B Antibodies cause foreign cells to clump together.

Your red blood cells can have antigens called A and B. If you have A antigens, you are **blood type** A, and you have antibodies *against* antigen B. So if cells with antigen B get into your body, they will be clumped together by the antibodies, causing a blood clot. If you have neither A nor B antigens, you are blood type O.

Blood type	Red blood cells	Antibodies present in plasma	Antigens present on cells
A		anti-B	A
B		anti-A	B
AB		none	A and B
O		anti-B and anti-B	none

C The different antigens and antibodies present in different blood types.

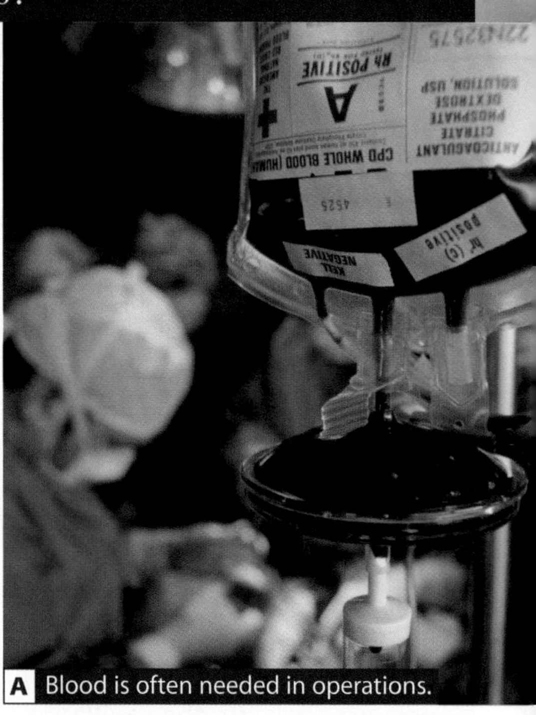

A Blood is often needed in operations.

?

1 What is plasma?

?

2 Which blood type are you if you have cells with B antigens?

3 Which blood type is the patient in photo A?

4 When someone is given blood, the antibodies in the new blood become very diluted so don't usually cause problems. With this in mind, what sorts of blood could people with the following blood types receive:
a type A
b type O
c type AB?

5 Why could giving type A blood to someone who is type B kill them?

H ABO alleles

There are three alleles (versions of genes) that control the antigens of your red blood cells: A, B and O. If you have the A allele, you have A antigens.

Your cells contain two copies of each chromosome and so have two copies of each gene. These copies may both be the same allele or different alleles. Some alleles are **dominant** and stop other alleles having an effect. Other alleles are **recessive** – their effects are not seen if the other allele is dominant.

The O allele causes no antigens to be produced but is recessive to both A and B, so its effects are seen only if the person has two copies of the O allele. Alleles A and B are **codominant**, which means that one allele does not stop the other from having its effect.

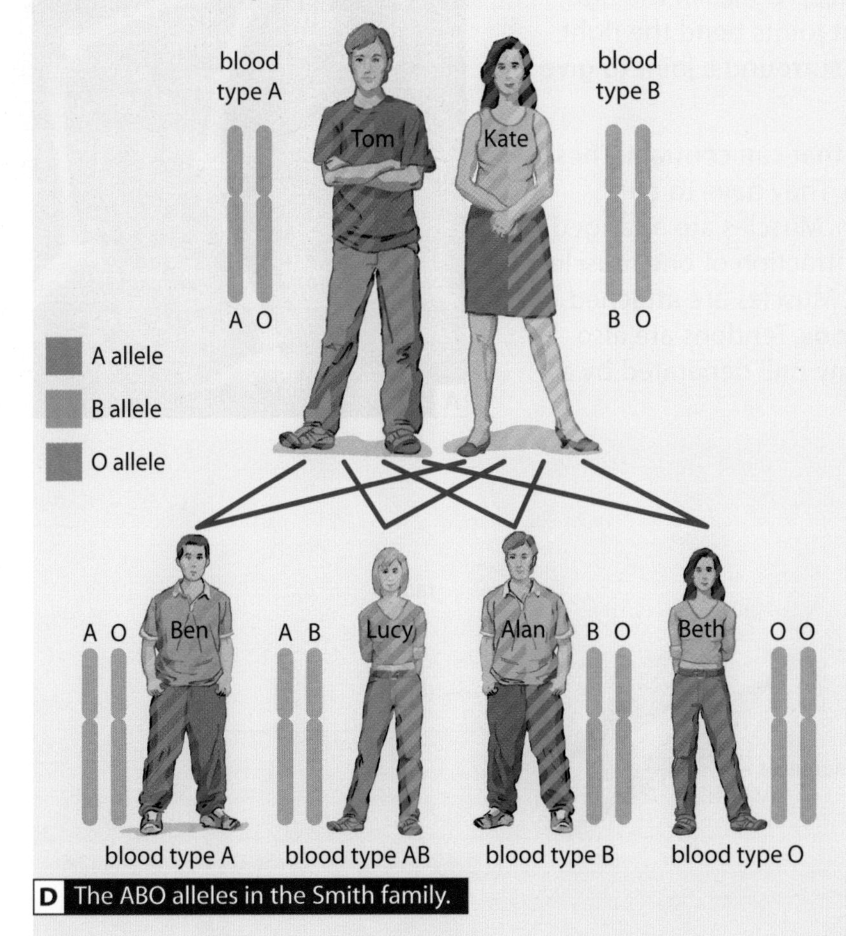

D The ABO alleles in the Smith family.

?

6 What antigens do you have if you have the B allele?

?

7 Look at drawing D.

- **a** Which alleles does Tom have?
- **b** Which of these alleles is dominant?
- **c** Which child has alleles that are codominant?
- **d** Which antigens does Beth have?

8 Punnett squares show the chances of children having a certain combination of alleles. For the Smiths there's a 1 in 4 chance (25%) for each combination. Copy and complete this Punnett square for the Smiths.

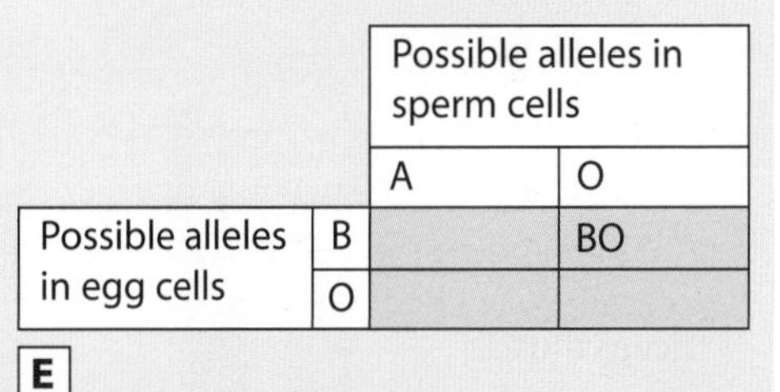

		Possible alleles in sperm cells	
		A	O
Possible alleles in egg cells	B		BO
	O		

E

9 **a** Draw a Punnett square for the children of Dave (alleles A and B) and Joan (BB).

b What are the percentage chances of a child getting each blood type?

Summary

Design a poster for a hospital to answer the question under the title on the previous page.

B7.20 Body structure

What holds us all together?

Humans, like all other vertebrates, have an internal bony skeleton. Some of the bones protect important soft and delicate organs. Other bones are **articulated**, so that we can move.

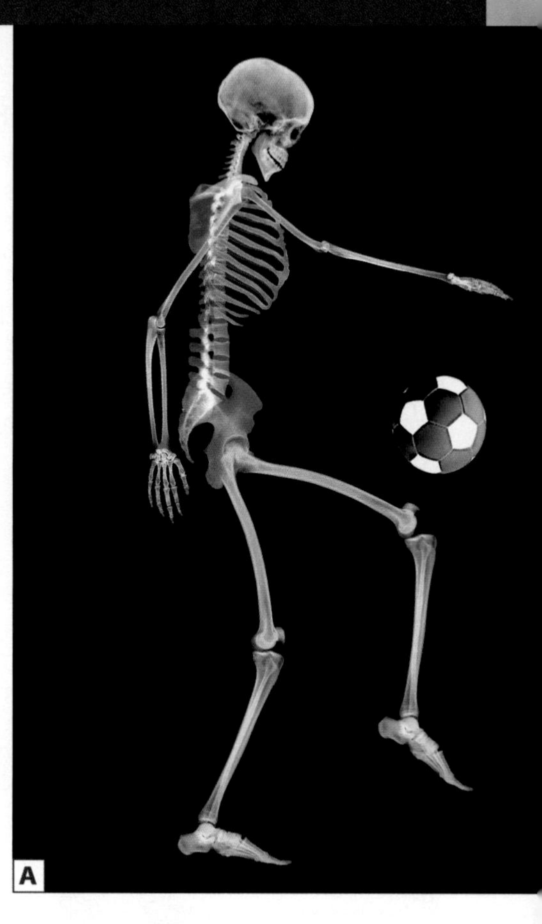
A

?

1. Name two organs in the human body that are protected by parts of the skeleton.
2. Explain what 'articulated' means, and why it is important for movement.

Bones are linked together by **ligaments**. Ligaments are made of tough cord-like tissue that keep bones in the right position relative to each other so that joints bend the right way. Ligaments may also completely surround a joint to give it added strength.

Muscles are made up of special cells that can contract. These cells can't extend again on their own. They have to be stretched back to their resting length. Muscles are arranged in **antagonistic pairs** so that the contraction of one muscle extends the other muscle of the pair. Muscles are attached to bones, and to each other, by **tendons**. Tendons are also tough and resist stretching, so that any pull generated by a muscle is transferred to the bone.

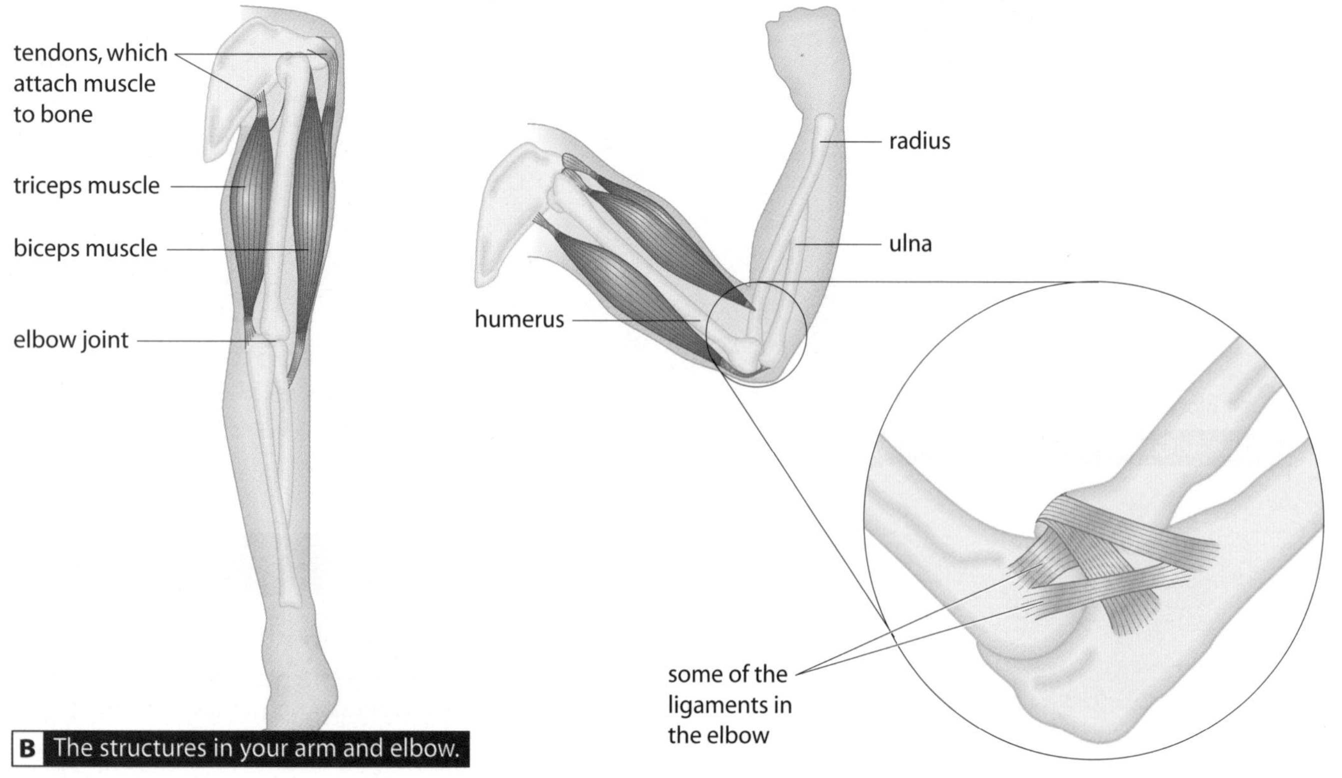

B The structures in your arm and elbow.

3 Describe one difference between ligaments and tendons.

4 Why are ligaments needed at joints?

5 a Name the muscles in one antagonistic pair.
  b Explain why our muscles are arranged in antagonistic pairs.

6 Why must tendons resist stretching?

The joints where bones meet and articulate are complex structures. The ends of the bones are covered in **cartilage**. Cartilage is smooth and slides more easily against another surface than bone does. It also absorbs shocks better than bone.

Many joints are enclosed in a **synovial capsule** that is made of tough, fibrous ligament tissue. The job of the capsule is to protect the structures inside and to contain **synovial fluid**. This fluid is a viscous oily liquid that lubricates the joint, making movement easier. The fluid also provides nutrients to the cartilage, because cartilage doesn't contain any blood capillaries.

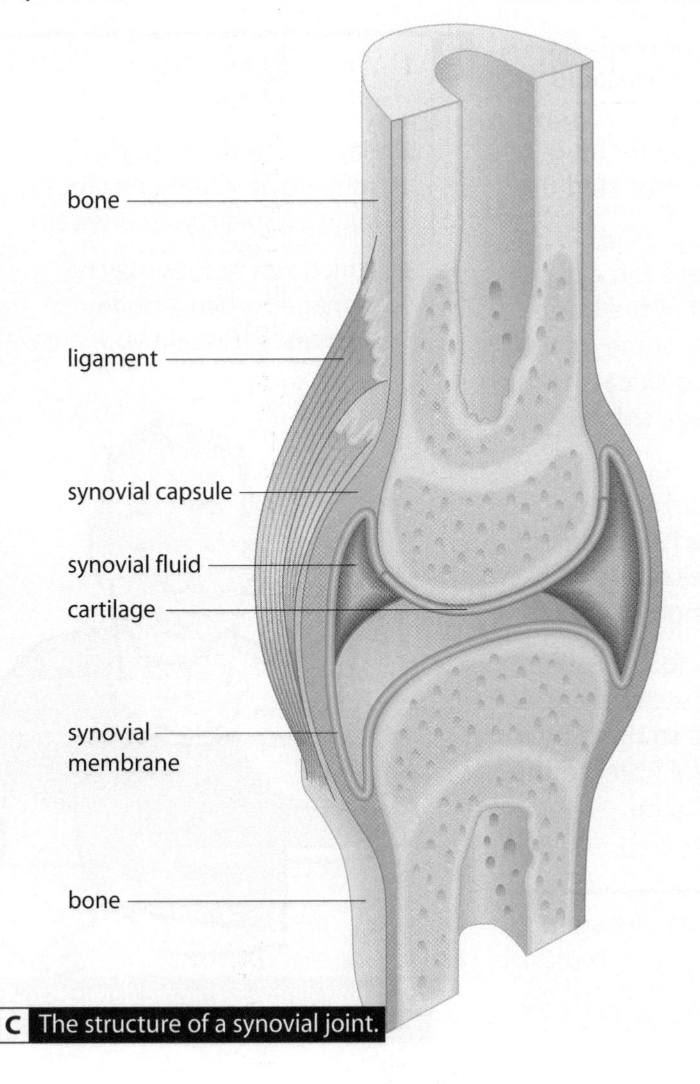

C The structure of a synovial joint.

7 Why are many joints enclosed in a capsule?

8 Explain the importance of cartilage at the ends of bones. Use the word friction in your answer.

9 Give two reasons why the cartilage at the ends of bones needs to be bathed in synovial fluid.

10 Synovial fluid is a lubricant, like oil in an engine. Explain how lubricants reduce friction.

Summary

Within one foot you have 26 bones, about 20 muscles, 33 joints and over 100 ligaments. Explain how the properties of the materials in these tissues make it possible for you to stand and walk.

B7.21 Injury from exercise

What kinds of injuries can you get from exercise?

A An awkward move can cause injury.

You will injure any tissue in your body if you apply a force that is beyond what it can normally cope with. This could happen during sport, in fitness exercise or just in everyday activities. **Acute** injuries happen when you do something suddenly. **Chronic** injuries may happen due to repeated use or overuse.

Bones may break, or **fracture**, if a large sideways force is applied. **Dislocation** of a bone, where the end of the bone moves out of place in a joint, is more common. Some joints, such as the shoulder and those in the fingers, dislocate most easily. During dislocation, the bone may damage soft tissues that surround the joint.

Ligaments are damaged when a joint is forced beyond its natural range of movement. This damage is called a **sprain**. The fibres of the ligament tear, causing pain and swelling. Large tears will weaken the joint, making dislocation more likely.

Tendons are most easily damaged where they join the muscle, because this is their weakest point. Acute damage to a tendon usually happens when a muscle is strained by a large or sudden force. This results in tears in the fibres in the tendon, which may be painful. Tendons can also be **inflamed** by repetitive use or overuse.

?

4 Explain why ligaments and tendons are easily damaged by overstretching.

5 Suggest how the man in cartoon B may have got his injury.

?

1 Why can injury happen during normal activity?

2 Has the person in photo A suffered an acute or a chronic injury? Explain your answer.

3 Which soft tissues may be damaged when a bone dislocates? Explain your answer.

B Tennis elbow is chronic damage to a tendon.

First-aid treatment of these kinds of injuries should follow the **RICE** routine: Rest, Ice, Compression, Elevation. This will help protect the area of injury from further damage. In many cases a doctor is needed to identify what has been injured, how extensive the damage is, and what treatment is needed. Some injuries may require surgery.

C The RICE routine for treating injuries.

As the injury is starting to mend, it will need suitable exercise to encourage the healing process and to begin to recover the full range of movement and strength (known as **rehabilitation**). This process may take many months.

?

6 Give examples of first-aid and long-term treatment that may be needed for sports injuries.

7 How can a physiotherapist help an injury to heal?

8 Even small amounts of damage to tendons may take many weeks to heal completely. Find out why, and what the dangers are of exercising again too soon.

D A **physiotherapist** suggests exercises and monitors progress to help recovery from injury.

Summary

Design a poster for your sports hall to show the kinds of injury that can be caused by excessive exercise, and what first-aid should be applied if they occur.

B7.22

Health information

What is information about your health useful for?

When you first visit your doctor, you may be asked some general health questions and have some measurements taken. A fitness instructor will also ask you questions before you start a fitness programme.

Your doctor may ask whether you smoke or drink alcohol, and what kind of diet you eat. They may also measure your blood pressure, height and weight. All these data indicate your general state of health. If they are not within normal ranges, they are **risk factors** for problems that can develop later in life, such as high blood pressure and heart disease.

Patient's name	*Jenna Fairview*
Energy intake	*c. 10.5 MJ a day*
Fruit/veg	*1 or 2 portions a day*
Smoking	*10–15 a day*
Alcohol	*18–20 units a week*
Exercise	*walks c. 1 km each day, swims about once a month*

A Jenna's doctor soon realised that Jenna didn't lead a healthy lifestyle.

?

1 The recommended average energy intake for a woman of Jenna's age and activity is 8.1 MJ per day.
   a Explain why this value is an average.
   b Why is this value recommended?
   c Suggest what effect Jenna's energy intake is having on her body.

2 What other factors in Jenna's answers indicate that she could be healthier?

3 Suggest what the doctor might recommend Jenna do about the different factors.

A fitness instructor will ask about existing medical conditions, such as high blood pressure, diabetes, any previous injuries or existing medication. They may also measure blood pressure, heart rate, body mass and flexibility. This information will show them how fit you are now and what kind of exercises you should be able to do.

B This fitness instructor makes sure his client knows how to do the exercises properly.

Both a doctor and a fitness instructor will ask whether there is any family history of conditions such as diabetes or heart problems. These conditions may be linked to genes and could be inherited.

H

C Your doctor can refer to all your previous records when you visit.

All the information that the doctor or fitness instructor collects will be recorded and stored. These data help them decide what is needed next, and will be used as a **baseline** for measuring improvement. The data are also available to other members of the team. For example, nurses within a doctor's practice may see a patient for some check-ups and need to know what the doctor has prescribed. And when you move and change doctor, your medical records will be sent to your new doctor.

D Annie discusses her clients with Jake so that he can monitor them while she is on holiday.

?

4 Why can blood pressure and heart rate indicate how fit or healthy you are?

5 Why are you asked about family history of some conditions, even if you don't show symptoms of those conditions?

?

6 Why is a doctor unlikely to ask about your previous medical history?

7 Give at least two reasons why information is stored and shared between members of the same health or fitness team.

8 Explain the importance of keeping detailed and accurate records.

9 Doctors are not usually allowed to share information about a patient with anyone outside the health team that is treating them. Find out why.

Summary

Draw a concept map to show information a doctor or fitness instructor collects to help assess how healthy you are. Include on your map why this information is useful to them.

H Add reasons to your map why this information needs to be stored and shared within the team.

B7.23 Monitoring fitness training

Why is monitoring during a fitness programme important?

A Different people have different reasons for going to a gym.

When the fitness instructor assesses a new client's health and fitness, they will also discuss what the client wants to achieve.

Gyms have a range of equipment for cardiovascular exercise (to improve the heart and circulation) and for weight training. They may also have a swimming pool and provide other facilities, such as exercise classes. This gives the instructor a range of choices that could be used to achieve what the client wants. The instructor will discuss this range with the client and choose what will help to achieve the agreed target and keep the client motivated to exercise.

C Some people enjoy the social side of exercising together.

?

1. Give as many different reasons as you can why people go to the gym.
2. Explain why the people in figure A need different fitness programmes. Give as much detail as possible.

B Gyms have a variety of equipment.

?

3. Explain why a fitness instructor will help a client choose which fitness programme to follow.
4. Explain why motivation is important in exercise and why it may vary between people.

During the fitness programme the instructor will need to see the client several times to take further measurements. These measurements will need to be recorded accurately so that they show what progress the client is making towards the target. This will also help to show whether the client is struggling with a particular activity, is at risk of damage from an exercise, or may be losing interest. In any of these cases, it is essential that the instructor discusses what is happening with the client and whether changes are needed to the fitness programme to help the client make better and safer progress.

D Monitoring of progress helps to improve performance.

?

5 Why is accurate record-keeping before and during a fitness programme essential?

6 Give as many reasons as you can why a fitness programme may be modified before it is completed.

H Measurements taken during the initial assessment, such as endurance, strength and flexibility, are repeated during the fitness programme. These measurements are taken several times to ensure that they are as accurate as possible. Ideally, they are also taken at the same time of day because these factors may vary in the same person at different times of the day. Measurements need to be accurate, so that the measure of progress towards the client's target is reliable.

?

7 a Explain how measurements are made as accurate as possible.
b Why is this important?

8 Why are the same measurements taken during a fitness programme as those used in the initial assessment?

9 Explain why measurements of endurance, strength and flexibility may vary during the day.

Summary

Philip's fitness instructor has changed some of the activities in his fitness programme. Give as many reasons as you can why the instructor did this, and explain what part the fitness tests Philip had every month played in deciding what needed to be changed.

B7.24 Treating arthritis

Why is monitoring during medical treatment important?

Arthritis is a medical term for inflammation of the joints. There many types of arthritis, each with different causes and different treatments.

Severe arthritis may be obvious, but a doctor may suspect mild arthritis from the symptoms that the patient describes. To check whether the patient has arthritis, and which kind it is, the doctor will examine the joint carefully and may have X-rays taken.

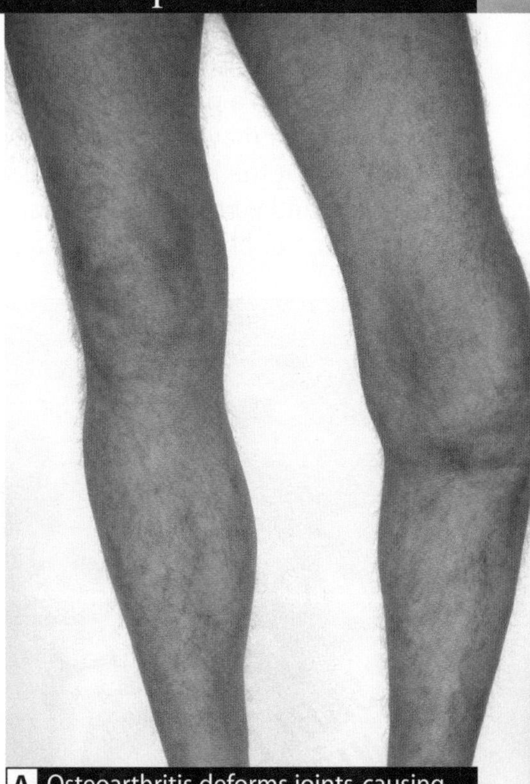

A Osteoarthritis deforms joints, causing pain.

?

1 a Which symptoms would suggest to a doctor that someone is suffering from osteoarthritis? Explain your answers.

b Suggest why these symptoms do not prove that someone has osteoarthritis.

2 How would taking an X-ray show whether or not a patient has osteoarthritis?

The different treatments for arthritis attempt to tackle different problems caused by the condition. Pain-killing drugs are prescribed to control pain. For mild cases, drugs such as aspirin or paracetamol may be used, but for more severe cases stronger pain-killing drugs are needed.

Many drugs have **side-effects** that are unpleasant. For example, aspirin causes stomach problems in some people, and steroids can cause bone thinning if taken over a long time. The doctor chooses the medication that gives the best pain control with the fewest side-effects for each patient. During treatment, the doctor will need to monitor the effectiveness of the medication. For example, changes may be needed if the condition improves or worsens, or if the patient reacts badly to the drugs.

B Cortisone is a powerful drug that is injected to control extreme pain in joints including the shoulder. It can cause many side-effects.

?

3 Explain why different patients with the same kind of arthritis may receive different medication.

4 How does a doctor decide which medication to prescribe?

5 Explain why a doctor may modify a patient's treatment.

The patient may also see a physiotherapist for massage or acupuncture to reduce pain. A physiotherapist can also suggest suitable exercises to improve mobility, and to reduce stress and muscle tension. Regular exercise helps to reduce weight and improve movement of the joint.

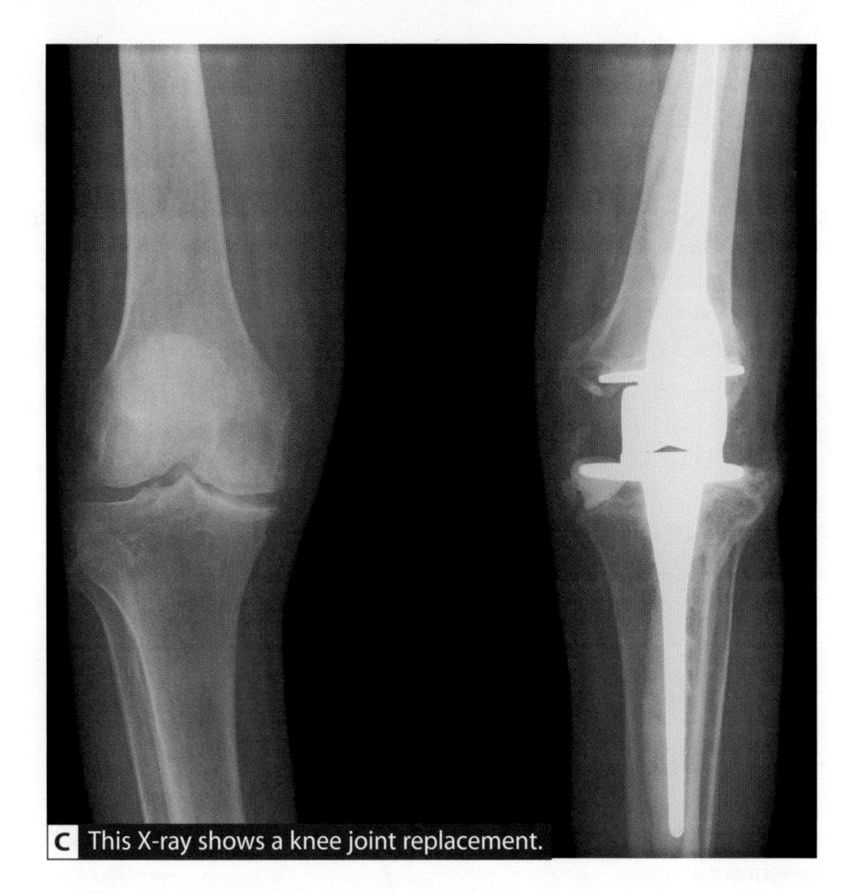

C This X-ray shows a knee joint replacement.

In extreme cases, the damaged joint may be replaced. Hip, knee and finger joint replacements are very common. When the patient returns home after an operation, the doctor will visit to monitor progress. Records will also be made by the community nurses who visit while the patient is recovering. A physiotherapist may also visit to help with rehabilitation of the joint.

?

6 Explain why accurate record-keeping is essential during a patient's treatment and recovery from an operation.

H Data collected about a patient include the patient's own recollection of pain and problems, the results of the doctor's examinations of the patient and any X-rays. Depending on the drugs taken, blood and urine tests may be done to check for side-effects such as reduction of numbers of blood cells or leakage of protein into the urine.

?

7 Comment on the accuracy of each form of monitoring during a patient's treatment for arthritis.

8 An operation to replace a damaged joint is carried out only when other treatments are unsuccessful. Explain why.

Summary

Make notes to answer the question at the top of page 56.

H Consider the monitoring techniques used, and explain how these affect the reliability of the data recorded.

Revision questions

1 The diagrams show the particles in a sandy soil and in a clay soil.

A

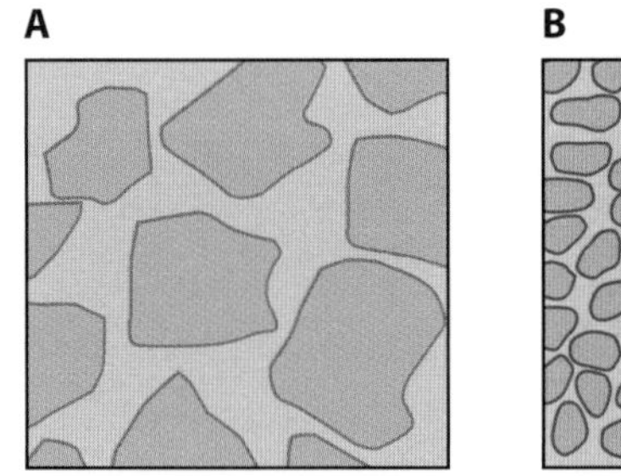

B

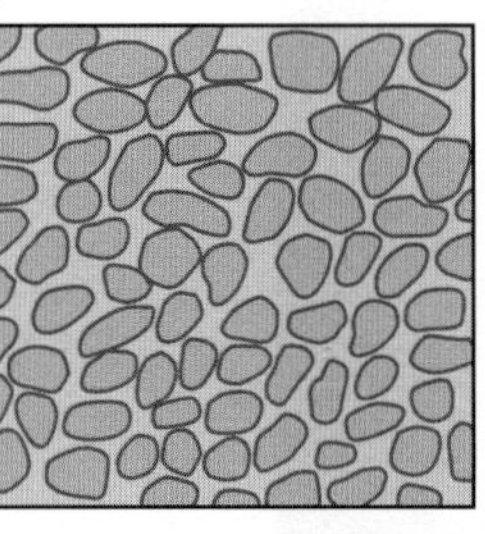

- **a** Which diagram shows a sandy soil? Explain your answer.
- **b** The diagrams show the inorganic particles in the soils. Give **three** possible sources of these particles.
- **c** If you took samples of each of these soils 24 hours after the last rainfall, and measured the percentage water content, which would you expect to have the higher water content? Explain your answer.

2 **a** Copy and complete this equation for photosynthesis:

water + _________ → _________ + oxygen

- **b** What **two** other things are needed for photosynthesis to occur?
- **c** What is the waste product of photosynthesis?

H **3** Much of the glucose made in photosynthesis is turned to starch.

- **a** Explain why starch is a better storage material than glucose.
- **b** Some glucose is converted into proteins. Which mineral ion is needed for this to happen?
- **c** How does this mineral ion get into the plant?

4 The graph shows the rate of photosynthesis for three different varieties of wheat.

- **a** Which variety of wheat grows best at lower temperatures?
- **b** How do you know this?
- **c** Explain why wheat is not grown in tropical climates.
- **d** Suggest how the rate of photosynthesis for KanKing wheat growing at 20 °C could be increased.

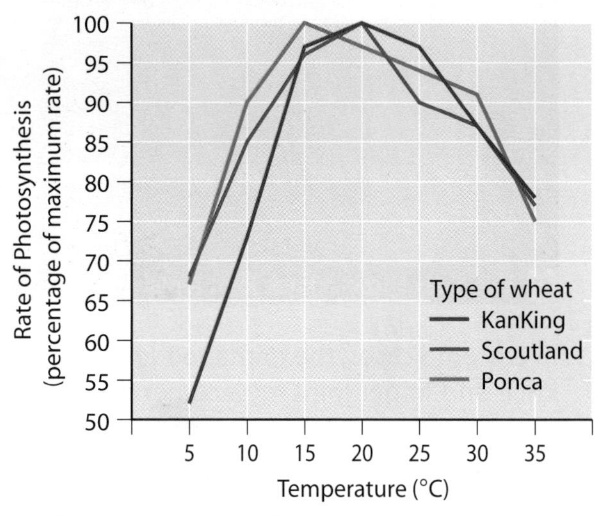

5 **a** Give an example of an animal parasite.

- **b** Explain how the parasite you have named is adapted to receive nutrition from its host.
- **c** Siboglinid tubeworms live at the bottom of deep oceans and have bacteria living inside them, which they feed on. The worms provide the bacteria with substances that they need in order to grow and respire. Is this an example of parasitism, symbiosis or commensalism?

H **6** Sickle-cell anaemia is common in areas where malaria is a problem and is caused by someone receiving two copies of a faulty allele. The disease normally results in death before the person is able to have children.

- **a** State **two** symptoms of sickle-cell anaemia.
- **b** Why would the sickle-cell allele disappear from a population where there was no malaria?
- **c** Explain why it has not disappeared in areas where there is malaria.

7 Genetic modification takes place in three steps.

a Put the following steps in order and explain what each means.

replication   isolation   transfer

b Which word best describes all the products obtained from genetically engineered bacteria?

chemicals   drugs   insulin   proteins

H

c The drawing shows a plasmid and a gene.

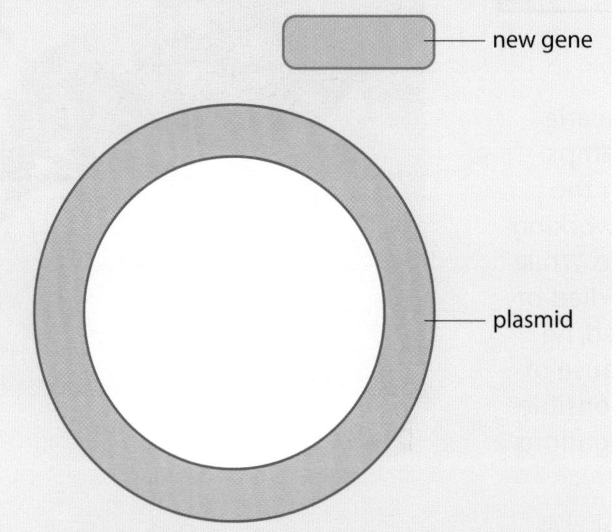

Make a series of drawings to show how the plasmid can be used as a vector to genetically modify a bacterium.

8 Cris sat down and measured his heart rate. He counted 72 bpm (beats per minute). He waited for a minute and then counted again. This time it was 60 bpm.

a Give **two** different explanations for the change in Cris's heart rate.

b Hannah measured her heart rate after sitting still for a minute. She counted 66 bpm. Cris said that both he and Hannah had normal heart rates. Explain why Cris was correct.

c Hannah went for a brisk walk and measured her heart rate again. Do you think this was 60 bpm, 66 bpm or 92 bpm?

d Explain your answer to part **c** as fully as you can.

9 Ravi has a problem with his heart. The tendons supporting the valve separating the left atrium and the left ventricle have become weak. The valve keeps turning inside out. Explain why this would cause a problem and what symptoms Ravi might suffer.

10 The diagram shows some of the tissues in the human arm.

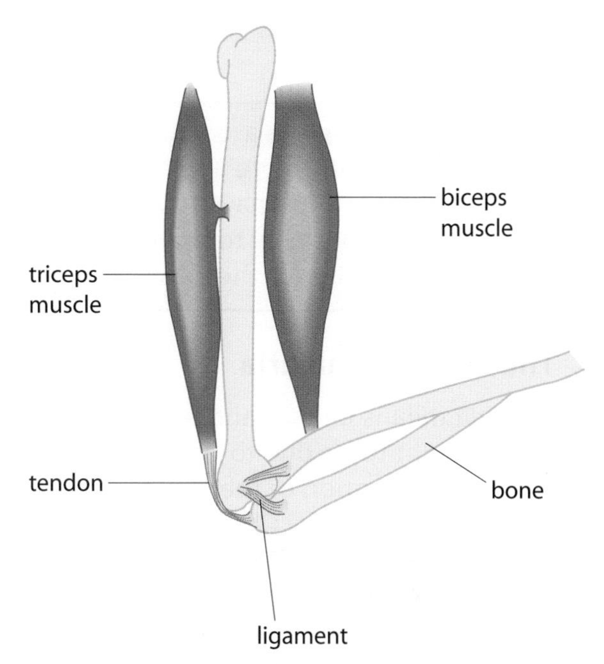

a What do tendons do?

b What properties must ligaments have? Explain your answer.

c Why are the ends of the bones covered in a layer of cartilage?

d The biceps and triceps muscles form an antagonistic pair.

i Why are muscles arranged in pairs like this?

ii Explain how this pair moves the bones of the lower arm.

Pre-release question

You will get a pre-release paper before the B7 exam. The text and questions here are to show you the kinds of questions you might have to answer, and to give you some hints about how to use the pre-release passage to help you to revise for the exam.

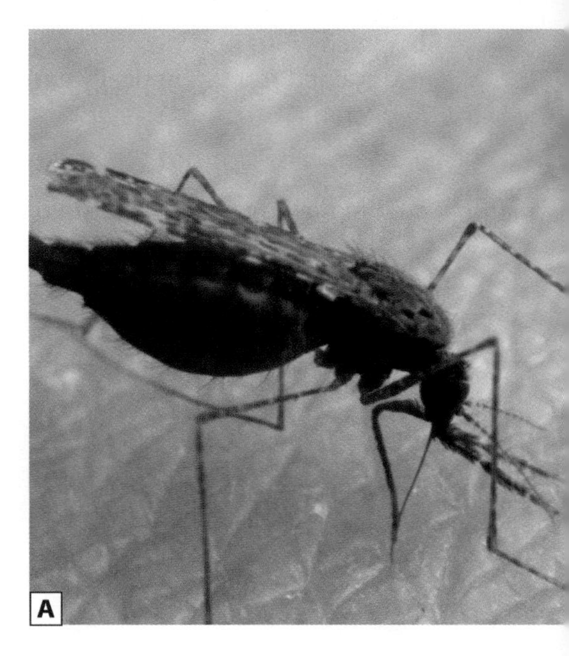
A

The cause of malaria

Mal aria means bad air, and refers to the fact that malaria was thought to be due to 'bad airs' coming from swamps. The first association of the *Plasmodium* parasite with the disease was made by Charles Laveran (1845–1922), working in Algeria as part of the French Army Medical Service. While there, he performed autopsies on patients who had died of malaria. In over three-quarters of the cases he studied, he found tiny parasites that he assumed must be the cause of malaria. He suggested that mosquitoes may be responsible for spreading the parasite, but did no further investigations to confirm this idea.

Ronald Ross (1857–1932) worked in the Indian Medical Service. While attending courses in England he became interested in malaria, and learned how to identify the parasites that caused the disease. When he returned to India, he started to investigate the theory that mosquitoes were responsible for spreading the disease. He eventually managed to identify *Plasmodium* parasites in the blood of *Anopheles* mosquitoes that had been allowed to bite malaria victims. At this point he switched his research to a bird form of malaria, and the *Culex fatigans* mosquito that carried it. He found the malaria parasite in the salivary glands of the mosquito, and he showed that healthy birds could be given malaria by being bitten by a mosquito that had fed on infected birds. This work was presented at a meeting of the British Medical Association.

Scientists in Italy then found similar results using the *Anopheles* mosquito and human patients.

When you get your pre-release paper, read it carefully and then think about which areas of science it covers. Go back to the module where this was covered and revise any of the content that looks relevant to the passage. The modules you need may be in the GCSE Science or the GCSE Additional Science book. Also think about the 'Ideas about Science' that might be relevant to the text (see the list on page 9).

The passage above is about malaria, which is a disease caused by parasites. This was covered in Module B7. Module B2 also looked at aspects of disease and health.

Some of the 'Ideas about Science' that could be asked about include how scientists develop explanations for observations, how scientific findings are reported and checked, and ethical considerations about experimental methods. Module B2 looked at the way new medicines are tested, so ideas in that module may be relevant to using humans to discover how a disease is transmitted.

If the passage above was given to you as your pre-release paper, it would be a very good idea to revise what you learned in B7, and also to take a look at B2 before the exam. This doesn't mean that you *will* get questions on these modules, just that you *might*.

The questions below are similar to the types of questions you may get in the exam. You will not see the questions before the exam, only the text.

1 Why was it important to discover the cause of malaria? [2 marks]

2 A swamp near a city was drained, and doctors noticed that the incidence of malaria in the city went down.
  a How could this be seen as evidence of the 'bad air' theory about the cause of malaria? [1 mark]
  b According to the current theory of how malaria is spread, why would draining the swamp reduce the incidence of the disease? [2 marks]

3 The *Plasmodium* parasite is a protozoan. Give the names of **two** other classes of organism that can cause diseases. [2 marks]

4 Laveran examined the blood of 192 patients when he was investigating the cause of malaria.
  a In approximately how many patients' blood did he find the parasite? [1 mark]
  b Laveran's discovery of *Plasmodium* in the blood of malaria patients suggested that the parasite caused the disease, but did not prove it. What further evidence could he have collected that would have made his case stronger? [2 marks]

5 Suggest why it was easier for Ross to work with malaria in birds rather than the human variety. [2 marks]

6 The Italian scientists should have obtained informed consent from their volunteers before conducting experiments on them.
  a What do you think *informed* consent means? [2 marks]
  b Why is it important? [1 mark]

GCSE Chemistry

GCSE Chemistry includes work from the GCSE Science book (Modules C1, C2 and C3) and from the GCSE Additional Science book (Modules C4, C5 and C6), as well as the work in this module.

The table shows how your GCSE Chemistry will be assessed. You may have already done the tests for Units 1 and 2, and possibly also some of the coursework.

Unit	Type of assessment	Tests you on ...
1	**Test paper** 40 minutes written paper (42 marks)	C1 Air quality C2 Material choices C3 Food matters
2	**Test paper** 40 minutes written paper (42 marks)	C4 Chemical patterns C5 Chemicals of the natural environment C6 Chemical synthesis
3	**Test paper with pre-release** 60 minutes written paper (55 marks)	C7 Further Chemistry Pre-release question (see below)
4	**Coursework** Practical Data Analysis (16 marks)	How well you can analyse and evaluate data from an experiment. See page 158.
	Case Study (40 marks)	How well you can gather and interpret information on a scientific subject, and present conclusions. See page 156.
5	**Coursework** Practical Investigation (40 marks)	How well you can plan and carry out a full investigation, and how well you can interpret your data and evaluate your data and conclusions. See page 158.

Pre-release question

The pre-release question is a passage of text, with questions based on the text or on science related to the text. The science involved could be any topic from Modules C1 to C7. You will be given the passage of text (but not the questions) before the examination, so you have time to look up the science. The questions may also test what you know about 'Ideas about Science'. You have been taught about these 'ideas' throughout the modules (in Biology and Physics modules as well as Chemistry). There is a practice question of this kind on page 110.

Ideas about Science

The Ideas about Science are:

- **Data and their limitations** – why results can vary or be in error, and how to make the best estimate of a true value.
- **Correlation and cause** – how to investigate the relationship between a factor and the outcome of an investigation, the difference between correlation and cause, and the need for a plausible mechanism linking a factor and an outcome.
- **Developing explanations** – how scientists explain things we observe, and how their explanations are tested.
- **The scientific community** – how scientists work together, and how they check and accept (or reject) scientific findings and explanations.
- **Risk** – what risks are, how we perceive them, and why some risks are accepted when others are not.
- **Making decisions about science and technology** – how some aspects of science raise ethical issues, and how decisions about science depend on social and economic factors as well as scientific factors.

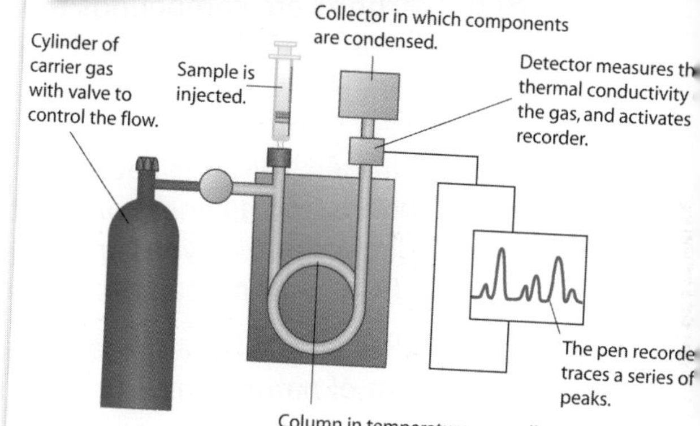

C7 Objectives

When you have studied Module C7 you should be able to:

- recall the structures and properties of the alkanes, alcohols, and carboxylic acids
- describe how esters are formed, and some of their properties and uses
- understand what activation energy is, and use energy level diagrams to represent exothermic and endothermic reactions
- understand what reversible reactions are, and how dynamic equilibrium can explain the differences between strong and weak acids
- describe the difference between qualitative and quantitative analysis, and explain the need for standard procedures
- describe how chromatography is carried out, and the differences between paper, thin-layer and gas chromatography
- explain what the mobile and stationary phases are, why locating agents are sometimes needed, and what retention time is
- describe how to carry out a titration, and analyse the results
- recall the differences between bulk and fine chemicals, and some examples of each
- describe and explain the characteristics of green chemistry
- compare the different methods of making ethanol.

C7.1 Alkanes

What are alkanes?

A All of these contain organic chemicals.

Organic compounds are compounds that contain carbon. The only common carbon compounds that are not called 'organic' are metal carbonates, carbon dioxide and carbon monoxide.

There are more than six million different organic chemicals, so scientists need to sort them into groups according to their properties. The groups of organic compounds studied in this module are hydrocarbons, alcohols, carboxylic acids and esters.

Hydrocarbons are compounds that contain carbon and hydrogen only. They are subdivided into alkanes, alkenes and alkynes. **Alkanes** contain carbon–carbon single bonds and are said to be saturated, as they cannot contain any more hydrogen. **Alkenes** have one or more carbon–carbon double bonds and alkynes contain carbon–carbon triple bonds. Alkenes and alkynes are said to be **unsaturated**, as more hydrogen atoms can be added to the molecule.

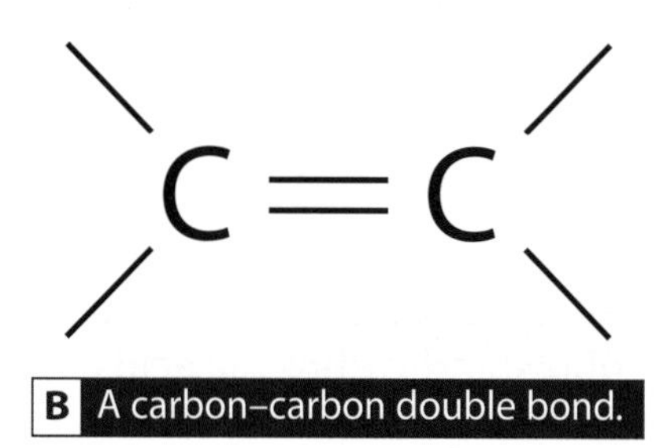

B A carbon–carbon double bond.

The simplest alkane is methane (CH_4), which is the main component of natural gas. Ethane (C_2H_6), propane (C_3H_8) and butane (C_4H_{10}) are also found in small amounts in natural gas, but are mainly obtained from the fractional distillation of crude oil. Table C lists these compounds with their **molecular formulae** and shows how their structures can be illustrated either as **structural formulae** or as **ball-and-stick models**.

Name of alkane	Molecular formula	Structural formula	Ball-and-stick model
Methane	CH_4	H H–C–H H	
Ethane	C_2H_6	H H H–C–C–H H H	
Propane	C_3H_8	H H H H–C–C–C–H H H H	
Butane	C_4H_{10}	H H H H H–C–C–C–C–H H H H H	

C

?

1. What is a hydrocarbon?
2. How can you tell whether a compound is a hydrocarbon by looking at its formula?
3. How do the formulae of alkanes change as the number of carbon atoms increases?
4. Predict the formula of pentane, which has five carbon atoms.

Alkanes are very useful fuels, burning in air to form carbon dioxide and water. If insufficient air is present, toxic carbon monoxide is formed and soot particles can be produced.

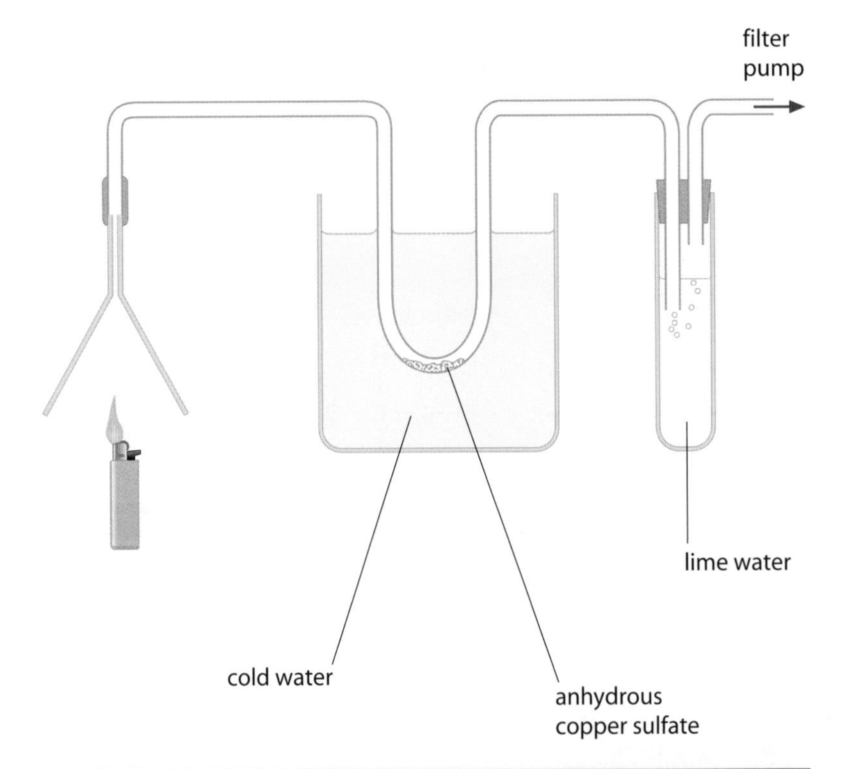

D The butane in this lighter burns to form carbon dioxide and water.

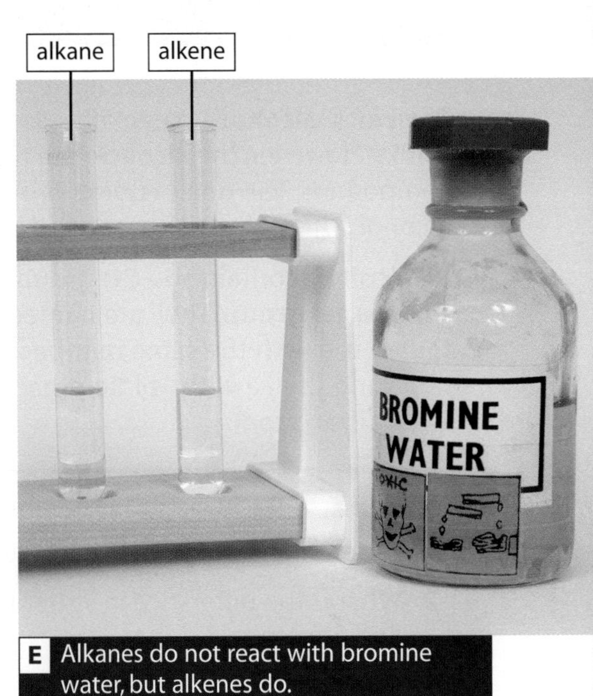

E Alkanes do not react with bromine water, but alkenes do.

Alkanes are unreactive because their carbon–carbon and carbon–hydrogen bonds are strong and therefore difficult to break. This can be shown by adding brown bromine water to an alkane; nothing happens. However, if an alkene containing a carbon–carbon double bond is mixed with bromine water, a reaction occurs and the bromine is decolorised.

?

5 **a** What is the difference between the structure of an alkane and that of an alkene?
b What is the difference in their reactivity?
c What do you think causes the reactivity of alkenes?

6 Alkenes also decolorise purple acidified potassium permanganate solution. What effect (if any) will an alkane have on this reactant? Explain your answer.

H All alkanes burn in oxygen to form carbon dioxide and water. As the temperature is high, we write $H_2O(g)$ for the water vapour formed.

The balanced symbol equation for burning methane is:

$$CH_4(g) + 2O_2(g) \rightarrow CO_2(g) + 2H_2O(g)$$

Ethane burns to give the same products:

$$2C_2H_6(g) + 7O_2(g) \rightarrow 4CO_2(g) + 6H_2O(g)$$

?

7 Write balanced symbol equations to show what happens when these compounds burn in plenty of air:
a propane
b butane.

8 Write balanced symbol equations for the burning of propane and butane in insufficient air to form carbon monoxide.

Summary

Write an encyclopaedia article with the title 'Alkanes'.

C7.2 Alcohols

Why are alcohols important?

The word '**alcohol**' is normally associated with alcoholic drinks. However, the alcohols are a whole group of organic compounds. The one in drinks is ethanol, C_2H_5OH. The alcohol with only one carbon is methanol, CH_3OH.

All alcohols contain an –OH group. This is called the **functional group**. They are named by taking the name of the alkane with the same number of carbon atoms and replacing the –e with –ol. So ethan<u>e</u> and ethan<u>ol</u> both have two carbon atoms.

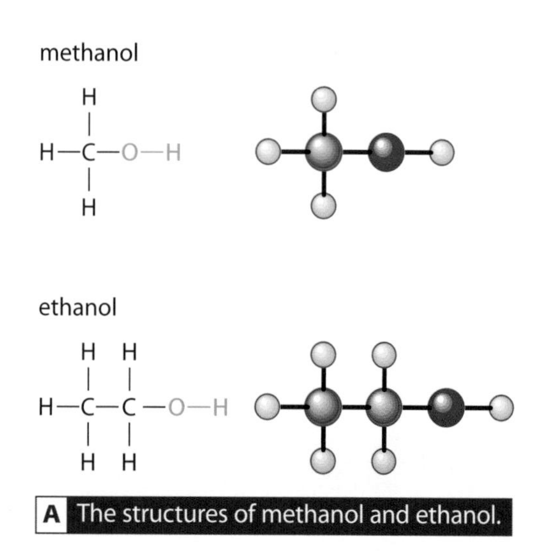

A The structures of methanol and ethanol.

?

1 Which of these are alcohols and which are alkanes?

propane, hexane, propanol, butanol, pentane

2 What is the difference in structure between an alkane and an alcohol?

3 Write the formula of propanol.

Methanol is important as a starting material for the plastics industry, and as a solvent. Pure ethanol is used as a fuel, especially in Brazil and the USA. Dilute ethanol produced by fermentation is used in drinks.

B Trangia stoves are designed to use meths as a fuel.

Methylated spirits ('meths') is a mixture of ethanol and methanol. Methanol is toxic and is added to meths to prevent people from drinking it. The purple colour is added to help people distinguish it from water, and also to make it taste foul.

Ethanol is a colourless liquid. Table C shows how it compares with water and ethane.

Name	Formula	Melting point (°C)	Boiling point (°C)
Ethane	C_2H_6	−183	−88
Ethanol	C_2H_5OH	−114	78
Water	H_2O	0	100

C

?

4 How do the melting and boiling points of ethanol compare with those of ethane?

5 Water can be written as H–OH. Suggest why it is being used as a comparison with ethanol.

D Brandy contains enough ethanol for it to burn.

All alcohols burn in air to form carbon dioxide and water. This is due to the hydrocarbon chain, just like alkanes burning.

The presence of the –OH functional group gives all alcohols similar properties, such as higher melting and boiling points than the equivalent alkanes. Alcohols can also be oxidised to form compounds called carboxylic acids. A bottle of wine left open to the air will become acidic as bacteria in the air oxidise the ethanol to ethanoic acid – also known as the acetic acid in vinegar.

?

6 Write a word equation showing how ethanol burns in air.

7 What would methanol form when it is oxidised?

H

Alcohols react slowly with sodium to produce hydrogen. For example:

ethanol + sodium → sodium ethoxide + hydrogen

$$2C_2H_5OH(l) + 2Na(s) \rightarrow 2C_2H_5ONa(s) + H_2(g)$$

This is useful way of getting rid of unwanted sodium, as the reaction is much less vigorous than the reaction between sodium and water:

water + sodium → sodium hydroxide + hydrogen

$$2H_2O(l) + 2Na(s) \rightarrow 2NaOH(aq) + H_2(g)$$

There is no reaction between ethane and sodium, as alkanes are so unreactive.

?

8 What group in ethanol must bring about the reaction between ethanol and sodium? Explain your answer.

9 Sodium ethoxide is an ionic compound. How could you show this?

Summary

Draw a poster showing all the reactions of ethanol.

Carboxylic acids

What are the properties of carboxylic acids?

Alcohols can be oxidised by bacteria in the air or chemically to form **carboxylic acids**. The two simplest carboxylic acids are methanoic acid, HCOOH, and ethanoic acid, CH_3COOH.

All carboxylic acids contain the –COOH functional group. It is this group that gives these acids their characteristic properties.

Many carboxylic acids have unpleasant tastes and smells. Methanoic acid, ethanoic acid and propanoic acid (C_2H_5COOH) all smell of vinegar. Butanoic acid is responsible for the smell of rancid butter and is present in human sweat.

methanoic acid

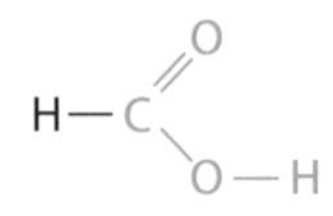

ethanoic acid

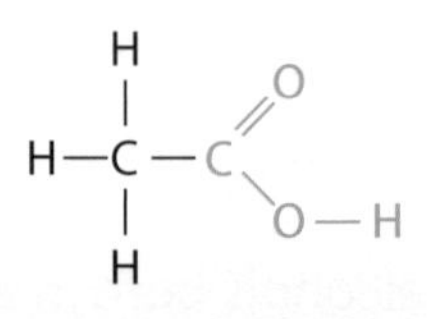

A The structural formulae of methanoic and ethanoic acids.

B Sweat contains a mix of carboxylic acids, each person's mixture being specific. Dogs are trained to recognise the smell of a person's sweat and so can track them.

?

1. How is the name of a carboxylic acid related to that of the alkane with the same number of carbon atoms?
2. Butanoic acid contains four carbon atoms. Write down its formula.

Carboxylic acids are weak acids, which means that they are only partially ionised in water. Their pH is about 2–3 and they will react more slowly than strong acids of the same concentration.

Carboxylic acids react with reactive metals like magnesium to form a salt and hydrogen. Methanoic acid forms methanoate salts, and ethanoic acid forms ethanoate salts:

ethanoic acid + magnesium → magnesium ethanoate + hydrogen

$$2CH_3COOH(aq) + Mg(s) \rightarrow (CH_3COO)_2Mg(aq) + H_2(g)$$

Carboxylic acids react with alkalis to form a salt and water:

methanoic acid + sodium hydroxide → sodium methanoate + water

$$HCOOH(aq) + NaOH(aq) \rightarrow HCOONa(aq) + H_2O(l)$$

Carboxylic acids react with metal carbonates to form a salt, carbon dioxide and water:

sodium carbonate + ethanoic acid → sodium ethanoate + carbon dioxide + water

$Na_2CO_3(aq) + 2CH_3COOH(aq) \rightarrow 2CH_3COONa(aq) + CO_2(g) + H_2O(l)$

The standard test for carboxylic acids is the production of bubbles of carbon dioxide when sodium carbonate solution is added to a carboxylic acid.

?

3 What are the products of the reaction between methanoic acid and zinc?

4 What salt is formed when potassium hydroxide reacts with propanoic acid?

5 How could you use sodium carbonate solution to distinguish between an alcohol and a carboxylic acid?

C All of these contain carboxylic acids.

Many carboxylic acids are found naturally. Vinegar is a dilute solution of ethanoic acid, and citric acid is found in all citrus fruits. Ascorbic acid (vitamin C) is a carboxylic acid found in citrus fruit and also in other fruit and vegetables. Ethanoic acid is also used to make acetate rayon, which is an artificial fibre.

?

6 Give two uses of vinegar.

7 Why do strong acids like hydrochloric acid react more rapidly than weak ones like ethanoic acid?

8 The salts formed are ionic. Remembering that a sodium ion is Na^+, work out the charge on an ethanoate ion.

Summary

Write a set of bullet points to go in the catalogue of a chemical supplier, explaining what a carboxylic acid is and summarising its reactions.

C7.4

Esters

Why do fruits smell sweet?

Carboxylic acids have reactions typical of any acid. They also react with alcohols to form compounds called **esters.** This reaction needs the presence of a strong acid such as sulfuric acid. The sulfuric acid acts as a **catalyst**, so it is not included in the equation. This reaction can be written in a general form:

carboxylic acid + alcohol → ester + water

This equation is similar to that showing the reaction of an acid with an alkali (which, like an alcohol, also contains an –OH group):

acid + alkali → salt + water

Esters all have very distinctive smells, and are responsible for the smells and flavours of fruits. Table B shows how the smell of an ester depends on the type of carboxylic acid and alcohol used.

Smell of ester	Carboxylic acid	Alcohol	Name of ester
banana	ethanoic acid	pentan-2-ol	pentyl ethanoate
peach	ethanoic acid	benzyl alcohol	benzyl ethanoate
pear	ethanoic acid	propan-1-ol	propyl ethanoate
pineapple	butanoic acid	ethanol	ethyl butanoate

B

'Pear drops' are flavoured with an ester called ethyl ethanoate, produced by the reaction of ethanoic acid with ethanol:

ethanoic acid + ethanol → ethyl ethanoate + water

C A kilogram of ripe pineapple may contain over 100 mg of ester.

A These sweets contain esters to give them their fruit flavours.

?

1 In what way are alcohols like alkalis?

2 Write a word equation for the reaction of methanoic acid with ethanol to form ethyl methanoate and water.

3 Write a word equation showing the formation of ethyl butanoate.

Esters are used mainly as flavourings and perfumes. They are also used as solvents, for example in polystyrene cement, and as plasticisers to help plastic remain flexible.

H Diagram D shows the apparatus used to make an ester. The carboxylic acid, alcohol and acid catalyst are placed in the flask. They are then heated under **reflux**. The carboxylic acid, alcohol and ester all have low boiling points, so they vaporise during the reaction. The condenser cools the vapours down so that they liquefy and fall back into the flask instead of escaping from the reaction mixture. When the reaction has finished, the ester is removed by distillation, as it has a different boiling point from the other organic compounds. The ester is then shaken with sodium carbonate solution in a tap funnel. This removes any acids that may still be present. The ester is then dried using calcium chloride and finally redistilled to purify it.

This is the symbol equation for the production of ethyl ethanoate:

$$CH_3COOH(l) + C_2H_5OH(l) \rightarrow CH_3COOC_2H_5(aq) + H_2O(l)$$

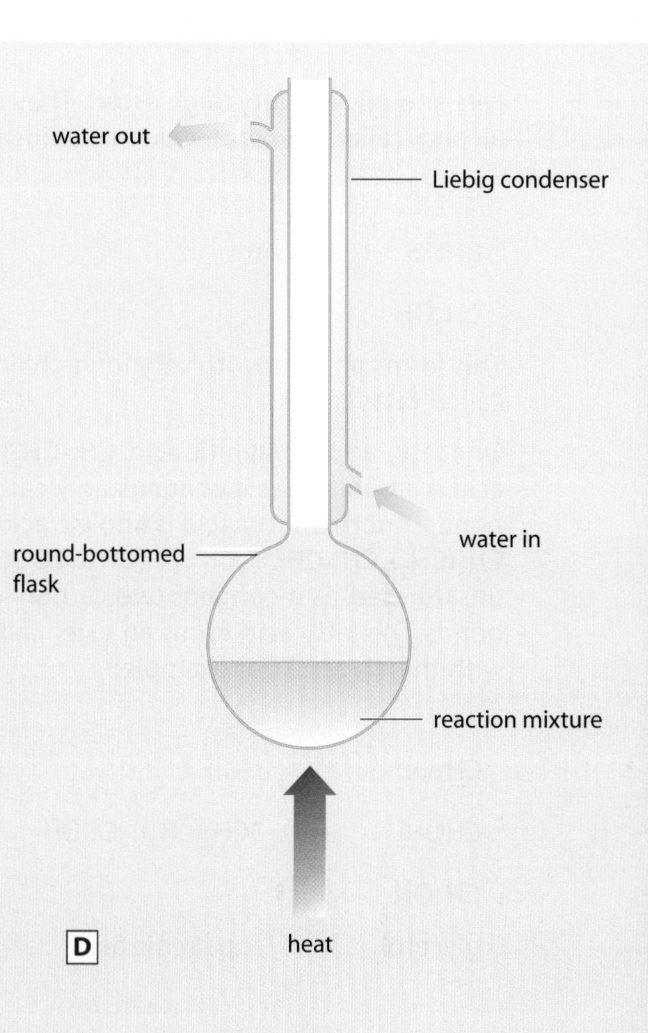

?

4 What does heating under reflux mean?

5 Ethanol boils at 78 °C, ethanoic acid at 118 °C and ethyl ethanoate at 77 °C. List the order in which they would distil off from a mixture.

6 **a** What acids might still be present with the ester?
b How are they removed?

7 Why are very pure esters needed for use in foods and perfumes?

8 Methyl ethanoate is CH_3COOCH_3 and propyl ethanoate is $CH_3COOC_3H_7$. Work out the general formula of esters.

Summary

Write a leaflet for a food company explaining to the public why they use esters and how esters are made.

C7.5 Fats and oils

What are fat and oils?

A Vegetable oils and animal fats.

Fats and oils are very large esters. They are made from an alcohol called **glycerol** which contains three –OH groups:

$$\begin{array}{ll} CH_2OH & \\ \vert & \\ CHOH & \text{glycerol} \\ \vert & \\ CH_2OH & \end{array}$$

This forms an ester with very long-chained carboxylic acids called **fatty acids**.

One fatty acid is palmitic acid, $CH_3(CH_2)_{14}COOH$. Palmitic acid is saturated, as it contains only carbon–carbon single bonds. Another fatty acid is linoleic acid, $CH_3(CH_2)_4CH{=}CHCH_2CH{=}CH(CH_2)_7COOH$. Linoleic acid is unsaturated, as it contains two carbon–carbon double bonds. The fatty acid forms an ester called a **triglyceride** with the glycerol. For example:

$$\begin{array}{ccccccc} CH_2OH & & & & CH_3(CH_2)_{14}COOCH_2 & & \\ \vert & & & & \vert & & \\ CHOH & + & 3CH_3(CH_2)_{14}COOH & \rightarrow & CH_3(CH_2)_{14}COOCH & + & 3H_2O \\ \vert & & & & \vert & & \\ CH_2OH & & & & CH_3(CH_2)_{14}COOCH_2 & & \\ \text{glycerol} & + & \text{palmitic acid} & \rightarrow & \text{glyceryl tripalmitate} & + & \text{water} \end{array}$$

Vegetable oils contain mostly unsaturated fatty acids, while animal fats contain mostly saturated fatty acids. This affects their properties, since the more carbon–carbon double bonds that are present, the lower the melting point. This is why unsaturated vegetable oils are liquid while saturated animal fats are solid.

?

1. Write a word equation for the formation of fats and oils.
2. Name the fat formed from glycerol and linoleic acid.
3. What is the difference between a saturated and an unsaturated fatty acid?
4. What is the molecular formula for palmitic acid?

Both fats and oils are important in our diet, as they provide an energy store. However, if you eat too much fat and do not exercise enough you can become obese. Table B shows the typical fat content of some foods.

Food	Amount of fat or oil (g/100 g of food)
carrot	0
banana	0.3
salmon	13
beefburger	17
cheddar cheese	34
butter	80
olive oil	100

B

?

5 Some fish, such as salmon, contain quite a lot of fat. Suggest why we are still encouraged to eat fish.

6 Which foods would you choose if you wanted to lose weight?

7 Does this table give a true indication of how much of each food we should eat? Explain your answer.

Unsaturated oils can be converted into saturated fats by reacting them with hydrogen in the presence of a nickel catalyst. For example, margarine is made from vegetable oils such as sunflower oil. The amount of hydrogen that reacts with the oil can be controlled. The more double bonds that are removed, the harder the margarine becomes.

C The origin of margarine.

D Using bromine water to see whether a fat or oil is saturated.

We can find out whether an oil, fat or fatty acid is saturated by adding brown bromine water. If the bromine turns colourless, unsaturated fats must be present. If it doesn't change colour, the fats must be saturated.

?

8 Why do unsaturated fats react with bromine water?

9 How could you use bromine water to compare the number of carbon–carbon double bonds in oils?

Summary

Explain the similarities and differences between fats and oils.

C7.6

Energy changes in reactions

Why are there energy changes during chemical reactions?

Chemical reactions are always linked with an energy change. Most reactions are **exothermic**. These give out heat energy to the surroundings, which get hotter. In an **endothermic** reaction, heat energy is taken in from the surroundings, which get colder. In many endothermic reactions continuous heat is needed for the reaction to work.

Exothermic and endothermic changes can be shown using energy level diagrams. These show the energy of the reactants and the products, and also the difference in energy, which is called ΔH (pronounced delta H). In an exothermic reaction, where energy is lost to the surroundings, ΔH is negative. In an endothermic reaction, where energy is gained, ΔH is positive.

A The reaction between caesium and water is exothermic.

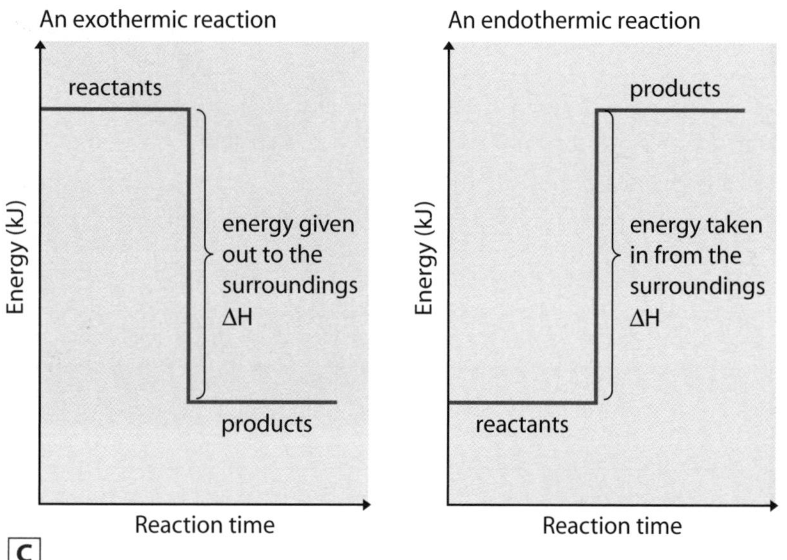

C

When chemicals react together, bonds between the atoms of the reactants must be broken. This needs energy. The atoms then rearrange themselves to form the products and new bonds are formed, giving out energy. The amount of energy needed to break the bonds at the start of the reaction is called the **activation energy**. Diagram D shows how this can be represented on energy level diagrams.

B Heat has to be taken in to make copper carbonate decompose into copper oxide and carbon dioxide.

?

1. Give two examples of an exothermic reaction.
2. Give two examples of an endothermic reaction.
3. Draw energy level diagrams for the reactions shown in photos A and B.

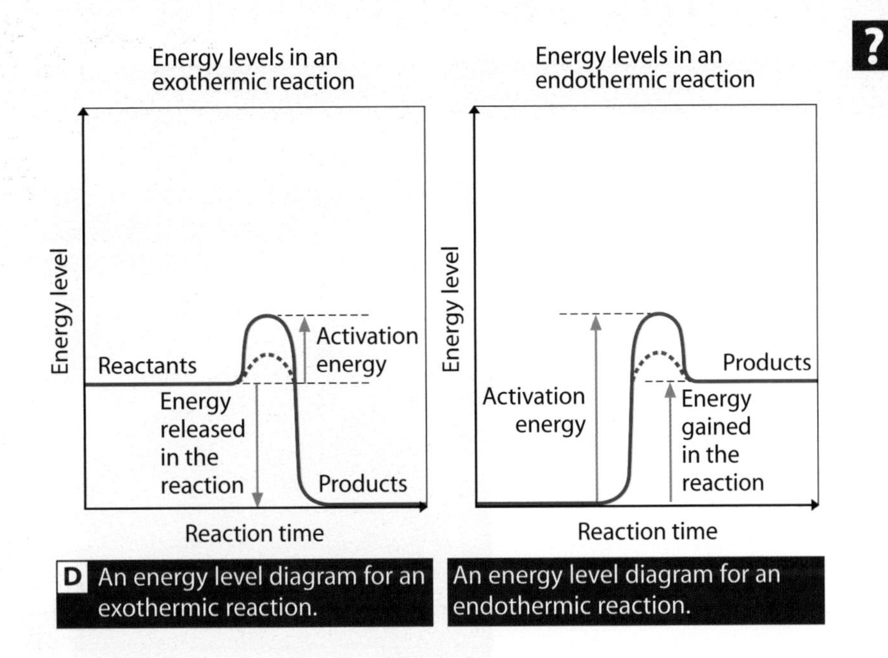

D An energy level diagram for an exothermic reaction.

An energy level diagram for an endothermic reaction.

?

4 Look at the left-hand graph in D.
   a How much energy is needed to break the bonds in the reaction?
   b How much energy is given out when the new bonds form?
   c What is the overall energy change during the reaction?

H Each covalent bond has its own specific amount of energy that is needed to break it. This is called the **bond energy**. The same amount of energy is given out when that particular bond forms. We can use bond energies to calculate the overall energy change in chemical reactions. Table E gives the bond energies of some common bonds.

Bond	Bond energy (kJ/mole)
C–C	347
C–H	413
C–O	358
C=O	805
Cl–Cl	243
H–H	436
H–Cl	432
O–H	464
O=O	498

E

For the reaction between chlorine and hydrogen to form hydrogen chloride:

$Cl_2 + H_2 \rightarrow 2HCl$

Cl–Cl + H–H → 2H–Cl

Bonds broken = Cl–Cl = 243 kJ
H–H = 436 kJ
Total energy taken in = 243 + 436 = 679 kJ

Bonds formed = 2 H–Cl = 2 × 432 kJ
Total energy given out = 864 kJ

Energy change (ΔH) = energy taken in – energy given out
= 679 kJ – 864 kJ
= –185 kJ

The negative sign shows that more energy is given out as bonds form than was taken in when the bonds were broken, so the reaction is exothermic.

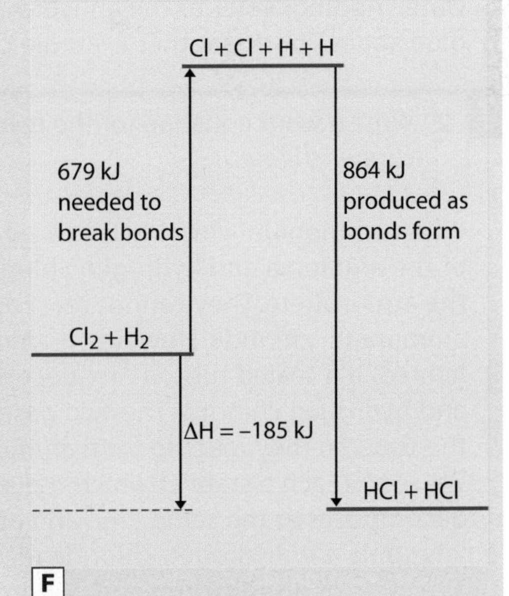

F

?

5 Calculate the energy change for the formation of water from hydrogen and oxygen: $2H_2 + O_2 \rightarrow 2H_2O$.

6 Why is the bond energy for a C=O bond larger than that for a C–O bond?

Summary

Explain why there are energy changes in chemical reactions.

Reversible reactions and equilibria

What is an equilibrium reaction?

In any chemical reaction, reactants are converted into products. However, in some chemical reactions, the products can react to form the original reactants. When you heat solid ammonium chloride, it decomposes to form gaseous ammonia and hydrogen chloride:

$$NH_4Cl(s) \rightarrow NH_3(g) + HCl(g)$$

If ammonia gas is held near hydrogen chloride gas, solid ammonium chloride is formed. The products of the first reaction have reacted to form the reactant again:

$$NH_3(g) + HCl(g) \rightarrow NH_4Cl(s)$$

This is called a **reversible reaction**. It can be written with arrows in both directions:

$$NH_4Cl(s) \rightleftarrows NH_3(g) + HCl(g)$$

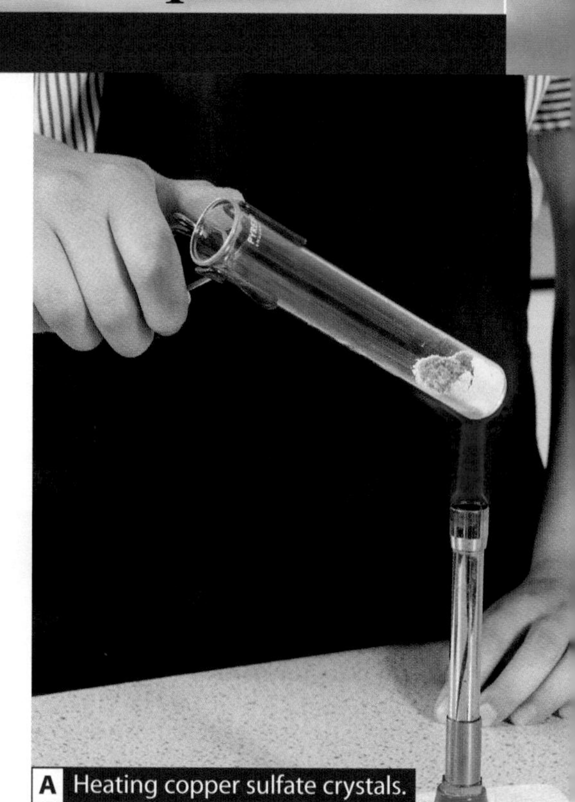

A Heating copper sulfate crystals.

?

1 How do you show that a reversible reaction is taking place when you write an equation?

When blue **hydrated** copper sulfate crystals are heated, they split up to form white **anhydrous** copper sulfate and water. Adding water to the anhydrous copper sulfate turns it blue again. This is another example of a reversible reaction.

?

2 Write a word equation for the reversible reaction involving copper sulfate.

When ammonium chloride is heated in an open test tube, all of the ammonia and hydrogen chloride formed escapes into the atmosphere. They cannot react again to form ammonium chloride. However, if ammonium chloride is heated in a sealed tube, it first decomposes into ammonia and hydrogen chloride. The two gases cannot escape from the tube, so they react to form ammonium chloride again. We soon reach a state at which the ammonium chloride decomposes at the same time and at the same rate as the ammonia and hydrogen chloride recombine. This is known as a state of **equilibrium** and is written $\rightleftharpoons$:

$$\text{ammonium chloride} \rightleftharpoons \text{ammonia} + \text{hydrogen chloride}$$

$$NH_4Cl(s) \rightleftharpoons NH_3(g) + HCl(g)$$

B Heating ammonium chloride in a sealed tube.

?

3 Why don't ammonia and hydrogen react again if they are produced in an open test tube?

When we talk about equilibrium reactions, we refer to the **forward reaction** and the **backward reaction**:

reactants $\rightleftharpoons$ products

forward reaction $\rightarrow$

$\leftarrow$ backward reaction

In a **dynamic equilibrium** the forward reaction and the backward reaction occur at the same time and at the same rate, so the amounts of each do not change.

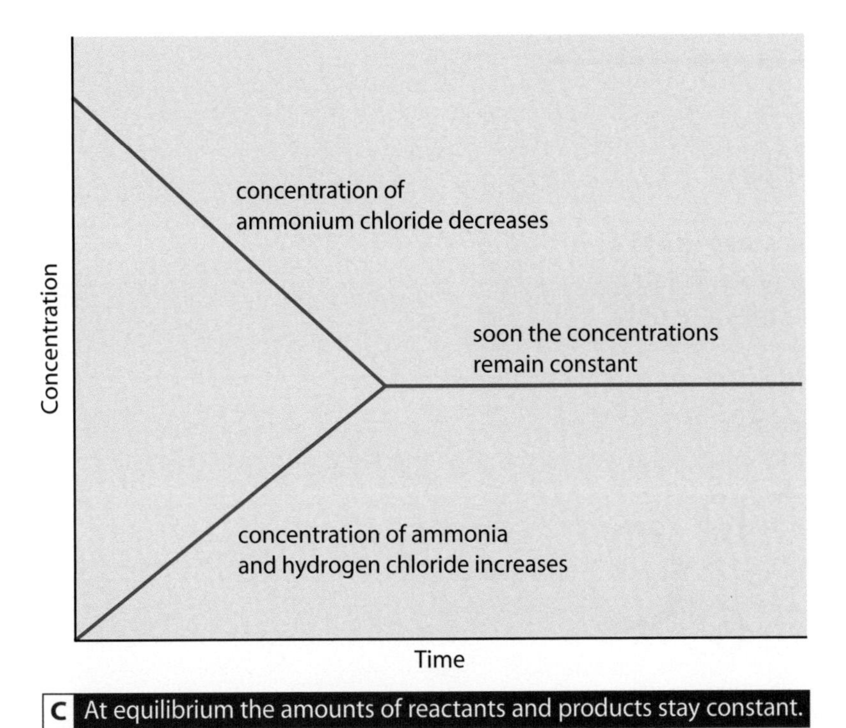

C At equilibrium the amounts of reactants and products stay constant.

Strong acids, like hydrochloric acid and nitric acid, are totally ionised in water. The forward reaction predominates and all of the acid splits up into ions:

$HCl(aq) \rightarrow H^+(aq) + Cl^-(aq)$

However, a **weak acid** like ethanoic acid (and all other carboxylic acids) is only partially ionised in water. There is a state of equilibrium between the non-ionised acid and the hydrogen ions and ethanoate ions formed:

$CH_3COOH(aq) \rightleftharpoons CH_3COO^-(aq) + H^+(aq)$

?

4 What is a state of dynamic equilibrium?

?

5 Write an equation showing what happens when nitric acid (HNO_3) totally ionises.

6 Write a symbol equation showing what happens when methanoic acid (HCOOH) ionises.

7 What is the difference between a reversible reaction and a reaction in equilibrium?

Summary

Explain, giving examples, what a 'state of equilibrium' means.

C7.8 Using equations to calculate amounts

How do we calculate amounts reacting together?

H We can use **balanced symbol equations** to represent chemical reactions. The physical state of each reactant or product is shown using state symbols. Remember that symbol equations must be *balanced*: there must be the same number of atoms of each element on each side of the equation, *without* changing any formulae.

?

1 Balance these equations:
   - **a** $C_3H_8(g) + O_2(g) \rightarrow CO_2(g) + H_2O(l)$
   - **b** $C_3H_7OH(l) + Na(s) \rightarrow C_3H_7ONa(s) + H_2(g)$

Each element has its own **relative atomic mass** (**RAM** or A_r for short). This is found using the Periodic Table. Relative atomic masses do not have units, because they are only *comparing* masses.

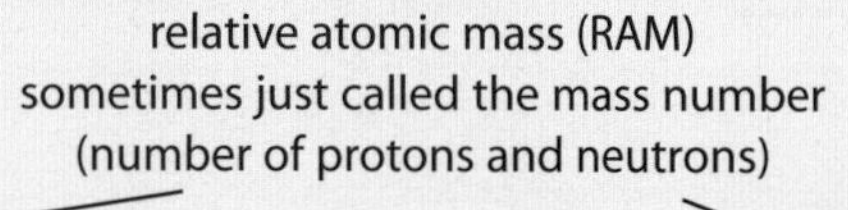

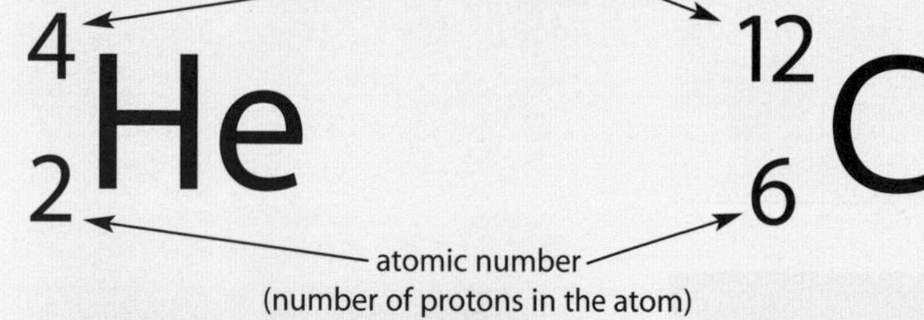

A Finding the RAM of an element.

The **relative formula mass** (**RFM** or M_r for short) is on the same scale as RAM, and again has no units. The RFM of a molecule can be found from its formula by adding up all the relative atomic masses of the atoms in the formula.

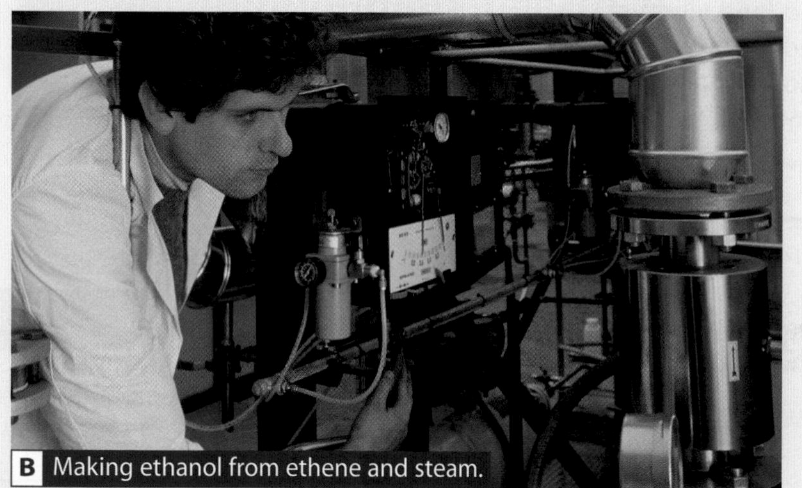

B Making ethanol from ethene and steam.

?

2 Use the Periodic Table on page 164 to find the RAMs of:
   - **a** arsenic (Group 5)
   - **b** xenon (Group 8).

3 Work out the relative formula masses of the following molecules:
   - **a** ethane, C_2H_6
   - **b** ethanol, C_2H_5OH
   - **c** ethanoic acid, CH_3COOH.

We can use relative formula masses to calculate the masses of reactants and products in a reaction.

Ethanol is produced industrially by the reaction of ethene with steam:

$$C_2H_4(g) + H_2O(g) \rightarrow C_2H_5OH(g)$$

How much ethene is needed to form 920 tonnes of ethanol?

Write out the balanced equation with the relative formula mass under each formula:

$$\underset{28}{C_2H_4(g)} + \underset{18}{H_2O(g)} \rightarrow \underset{46}{C_2H_5OH(g)}$$

This means that the ratio of the masses of ethene, steam and ethanol formed is 28:18:46, or that 28 g of ethene reacts with 18 g of steam to form 46 g of ethanol. Therefore we can scale up (× 20):

560 g ethene → 920 g ethanol

and change the units:

560 tonnes ethene → 920 tonnes ethanol

Ammonium chloride decomposes to form ammonia and hydrogen chloride:

$$NH_4Cl(s) \rightarrow NH_3(g) + HCl(g)$$

Find the mass of ammonia and hydrogen chloride formed when 214 g of ammonium chloride is heated.

$$\underset{53.5}{NH_4Cl(s)} \rightarrow \underset{17}{NH_3(g)} + \underset{36.5}{HCl(g)}$$

Therefore

53.5 g ammonium chloride
→ 17 g ammonia + 36.5 g hydrogen chloride

Scaling up (× 4):

214 g ammonium chloride
→ 68 g ammonia + 146 g hydrogen chloride

C Malachite contains copper carbonate.

?

4 Why is ethanol vapour formed when ethene reacts with steam?

5 46 g of ethanol produces 60 g of ethanoic acid.
How much ethanol is needed to form 300 kg of ethanoic acid?

6 a Copper carbonate decomposes to form copper oxide and carbon dioxide:
$CuCO_3(s) \rightarrow CuO(s) + CO_2(g)$
What mass of copper oxide is formed from 24.8 g of copper carbonate?

b 44 g of carbon dioxide occupies 24 dm^3 at room temperature.
What volume of carbon dioxide is formed in this reaction?

Summary

Explain how we calculate the amounts of reactants and products in a reaction.

Qualitative and quantitative analysis

What is the difference between qualitative and quantitative analysis?

Scientists often want to find out what is in unknown substances. **Qualitative analysis** is used to find out *what* is in a sample, for example the presence of a food additive in processed food, or the presence of blood at a crime scene. However, this type of analysis doesn't tell us the *amount* of substance being investigated – just whether it is present or not.

? 1 Why must we know whether certain chemicals are present in food?

A Limestone varies in its purity.

Quantitative analysis tells us *how much* of a particular substance there is in a sample. Before the use of modern technology, this was often done using chemical reactions. For example, limestone rocks are mainly calcium carbonate, but they also contain other substances. The calcium carbonate reacts with hydrochloric acid to form gaseous carbon dioxide:

$$CaCO_3(s) + 2HCl(aq) \rightarrow CaCl_2(aq) + CO_2(g) + H_2O(l)$$

The more calcium carbonate present in a sample of rock, the more carbon dioxide is produced and the greater the loss in mass.

Modern analytical methods are quick and accurate and can be used on tiny samples. For example, a scientist at a water company might need to find the amount of metal ions in a water sample. Many years ago, this was often done by **titration**, but this was slow and not always accurate. Modern analytical instruments depend on the use of electronics and computers. The sample of water is placed in one part of a machine, and a print-out showing the concentration of several different ions is produced quickly.

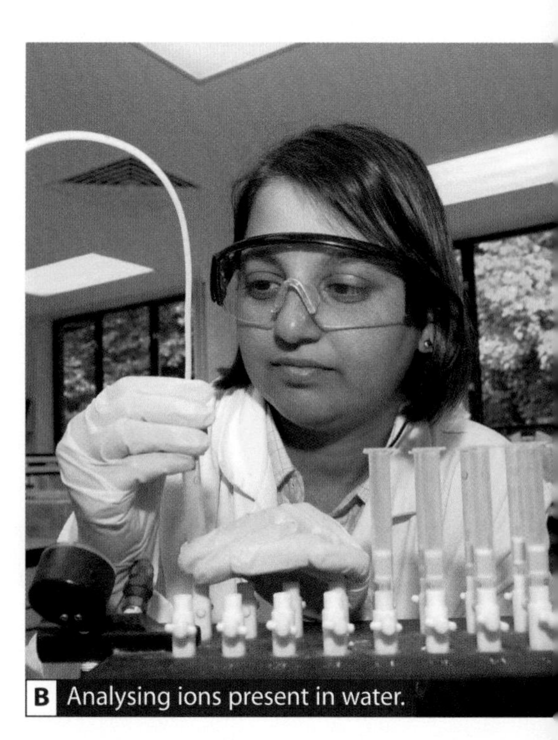

B Analysing ions present in water.

2 What is the difference between qualitative analysis and quantitative analysis?

3 Name two substances that a water analyst might be testing for.

4 Why is it important for analysis to be accurate and quick?

One of the problems that analysts may meet is that the sample to be analysed is not the same throughout. For example, two samples of limestone may have different compositions. A sample must therefore be used that represents the bulk of the material being analysed. For solids, a number of samples would be collected, crushed and then mixed together to form a more uniform sample.

If a sample dissolves in water, a solution is made and this is used for analysis. Here the reactants can mix together rapidly, and the concentration is the same throughout.

There must be a standard method for collecting, storing and preparing samples for analysis. When athletes have to produce urine samples for drug testing, there is a specified procedure. The sample is produced under supervision, and is then divided into two. One (sample A) is analysed at once, and the other (sample B) is stored in controlled conditions so that none of the chemicals that might be in the urine can change. If the first sample is positive, the second sample is analysed to make certain that the results are accurate.

5 Why must several samples be used for analysis?

6 What is the advantage of using aqueous samples?

7 What could happen if samples change during storage?

C This athlete was suspended from running, as analysis of his urine showed abnormal amounts of illegal substances in his body.

Summary

Describe the impact of modern technology on analysis.

Chromatography

How does chromatography work?

A Examples of coloured pigments.

The word 'chromatography' means 'colour writing', because the method was used first to separate coloured plant pigments.

You have probably used chromatography to study the pigments in grass, food colours or felt-tip pens. In each case the sample being studied was spotted onto some chromatography paper and then placed in a **solvent**. The solvent travelled up through the sample and separated it into a number of colours.

In chromatography, the chromatography paper is the **stationary phase** (as it doesn't move) and water or another solvent is the **mobile phase** (as it moves through the stationary phase). As the solvent passes over the sample some of the substances in the sample move into the mobile phase and get carried along. Other substances in the sample mixture are less soluble in the mobile phase and do not travel so far.

Water is said to be an **aqueous solvent**, while other solvents such as toluene and ethanol are called **non-aqueous solvents**. Non-aqueous solvents are used for chemicals that do not dissolve in water.

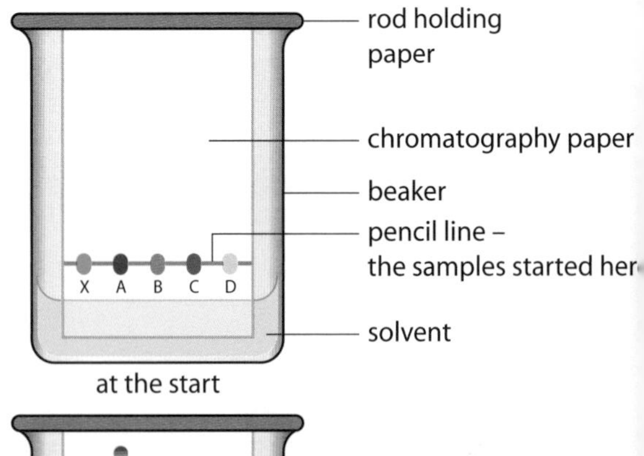

at the start

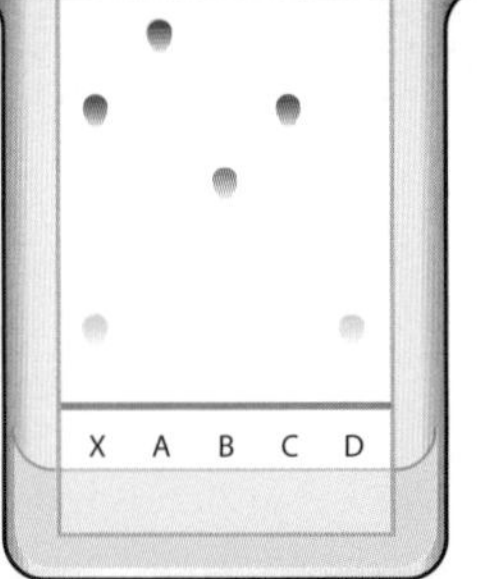

after the solvent has soaked up the paper

B Using chromatography to analyse a mixture.

?

1. What is the difference between the stationary phase and the mobile phase?
2. Look at diagram B.
   - **a** Which colour has travelled the furthest in the mobile phase?
   - **b** What dyes are in X?
3. What is the difference between aqueous and non-aqueous solvents?

Each of the chemicals in the sample being analysed is in a state of dynamic equilibrium between the stationary phase and the mobile phase:

chemical in stationary phase $\rightleftharpoons$ chemical in mobile phase

Some **solute** molecules in the sample are attracted more to the solvent, so the forward reaction dominates. These are the components that are carried along rapidly. Other molecules are attracted more to the stationary phase, so the backward reaction is more important. These are the components of the mixture that do not travel as far.

The choice of solvent for a particular separation is found by trial and error. The best solvent depends on the substances to be separated.

Chromatography can be used to identify substances in a mixture using **standard reference materials**. The same substance always moves at the same speed in the same solvent. A standard reference material is placed alongside the sample before chromatography. If the mixture contains the same substance as the standard, there will be a spot that has moved the same distance.

? **4** Look at diagram C. Which colour is most attracted to the stationary phase?

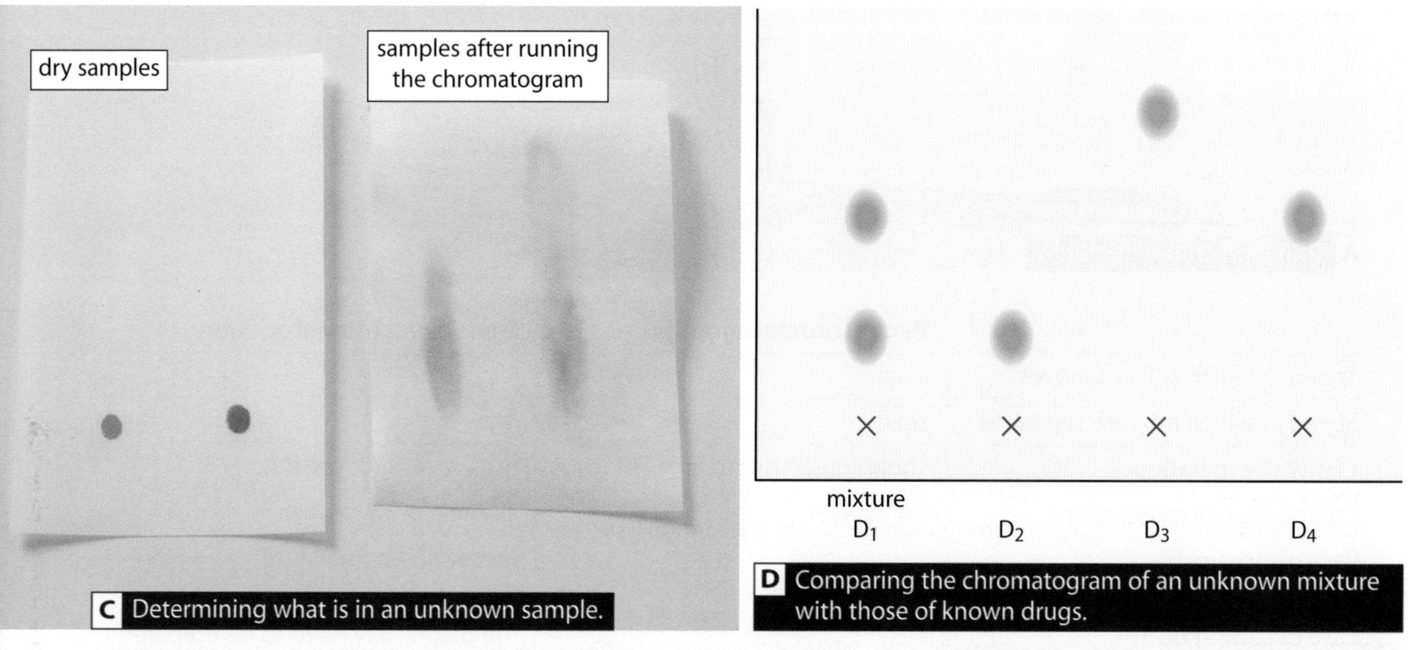

C Determining what is in an unknown sample.

D Comparing the chromatogram of an unknown mixture with those of known drugs.

This technique is used by forensic scientists to analyse samples of drugs, inks, paints and explosives. It is also used to test for food additives.

?

5 Diagram D shows a chromatogram of an unknown drugs mixture D_1 and three known drugs D_2, D_3 and D_4. Explain which of the known drugs are present in mixture D_1.

6 How could you show that each of the spots formed is pure?

Summary

Write a poster for a testing lab to use at a Food Show, explaining how they use chromatography to test for artificial colourings in food.

C7.11 Different types of chromatography

Why are there different types of chromatography?

Most chromatography carried out in schools is **paper chromatography**, where the stationary phase is specially treated filter paper.

An alternative is **thin-layer chromatography**, which uses a thin layer of a powder such as silica (silicon dioxide) or alumina (aluminium oxide) spread on a glass or plastic backing. This is used in an identical way to paper.

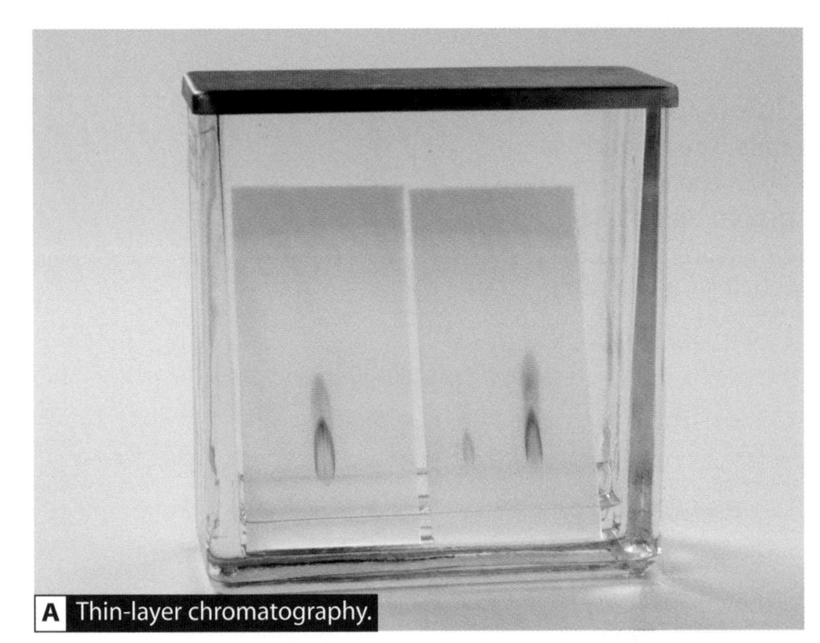

A Thin-layer chromatography.

	Paper chromatography	Thin-layer chromatography
Speed at which solvent moves	slow	fast
Speed at which mixture separates	slow	fast
Clarity of separation	spots spread out to form blobs	usually distinct spots
Choice of stationary phase	only paper	alumina or silica
Ease of storage of chromatograms produced	easy	these can be fragile
Ease of setting up	easy	the plates may need to be prepared first

B A comparison of paper and thin-layer chromatography.

Both paper and thin-layer chromatograms can be studied quantitatively. A pencil line is drawn to show where the original spot is. This is called the **origin**. The position that the solvent reaches at the end of running the chromatogram is called the **solvent front**, and must also be marked. The centre of each spot obtained is also marked.

?

1. Name two stationary phases that can be used in thin-layer chromatography.
2. Give three advantages and one disadvantage of thin-layer chromatography over paper chromatography.

The measurements shown in diagram C can be used to compare the distance moved by the substances in the mixture (*a*) to the distance the solvent front has travelled (*b*). The ratio is called the **retention factor** (**R_f**) value.

E

$$R_f = \frac{\text{distance travelled by the substance from the origin } (a)}{\text{distance travelled by the solvent from the origin } (b)}$$

?

3 Why are R_f values always less than 1.0?

4 What is the R_f of the spot in diagram C?

5 The pigments in green grass were separated by chromatography. Three spots were seen at 3.0 cm, 7.5 cm and 12.0 cm, when the solvent front had travelled 15.0 cm.
  a Draw the chromatogram to scale.
  b Calculate the R_f value for each spot.

Chromatography is also used nowadays to separate substances in mixtures that are colourless. There are different ways of detecting colourless spots.

Some substances are **fluorescent** when the chromatogram is viewed under an ultraviolet lamp.

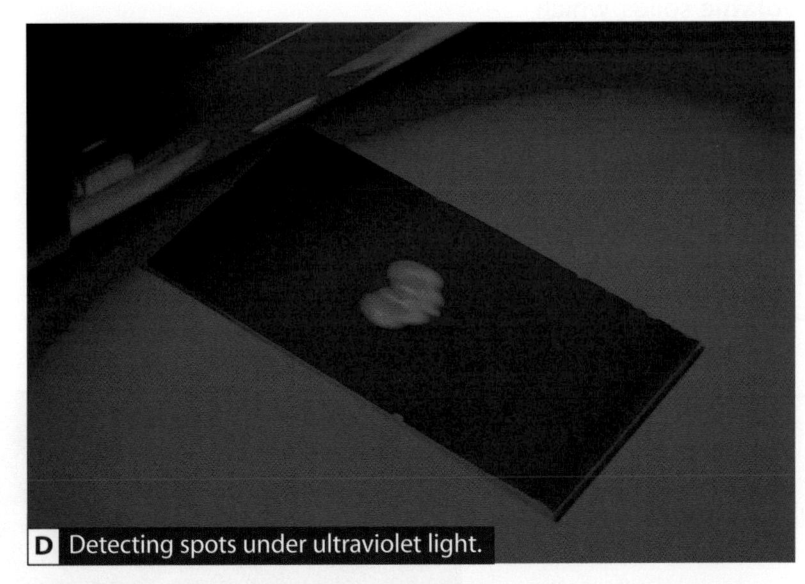

D Detecting spots under ultraviolet light.

Some colourless substances can be detected by spraying them with chemicals called **locating agents**, which react with the substances to form coloured spots. For example, amino acids are sprayed with ninhydrin solution, which produces purple spots when heated.

?

6 Why are locating agents needed?

7 There are 20 different amino acids. Some have identical R_f values in the same solvent. How could these be separated?

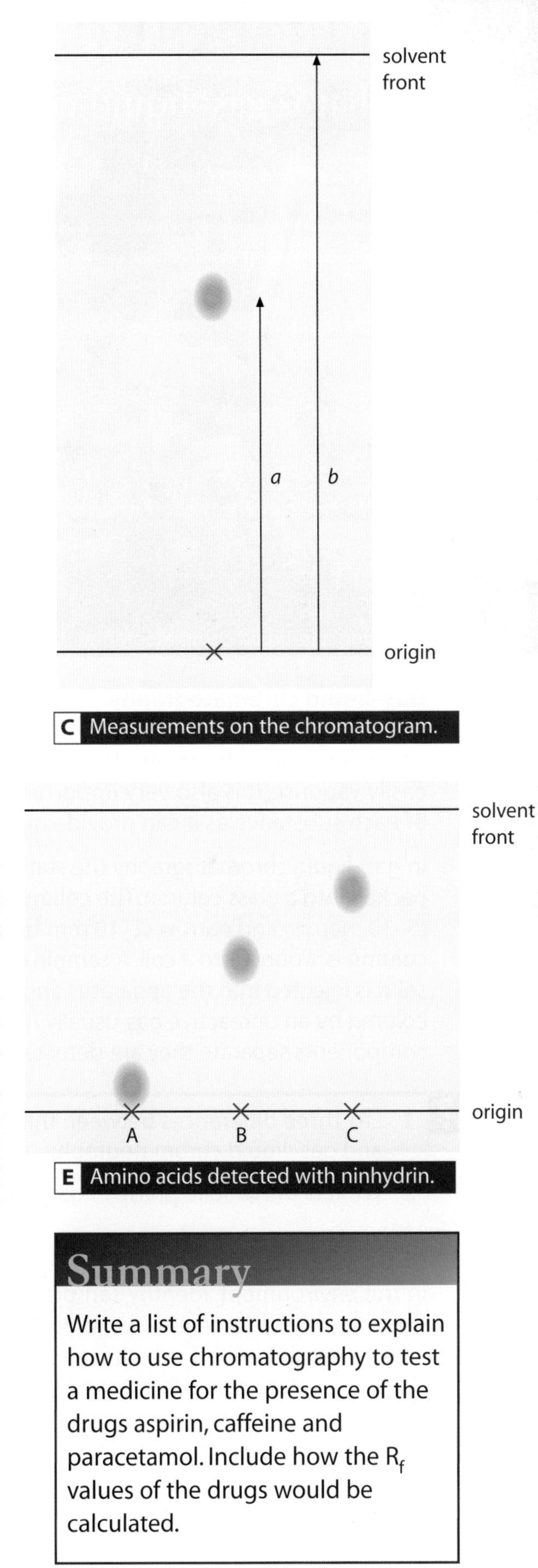

C Measurements on the chromatogram.

E Amino acids detected with ninhydrin.

Summary

Write a list of instructions to explain how to use chromatography to test a medicine for the presence of the drugs aspirin, caffeine and paracetamol. Include how the R_f values of the drugs would be calculated.

C7.12 Gas–liquid chromatography

What is gas–liquid chromatography used for?

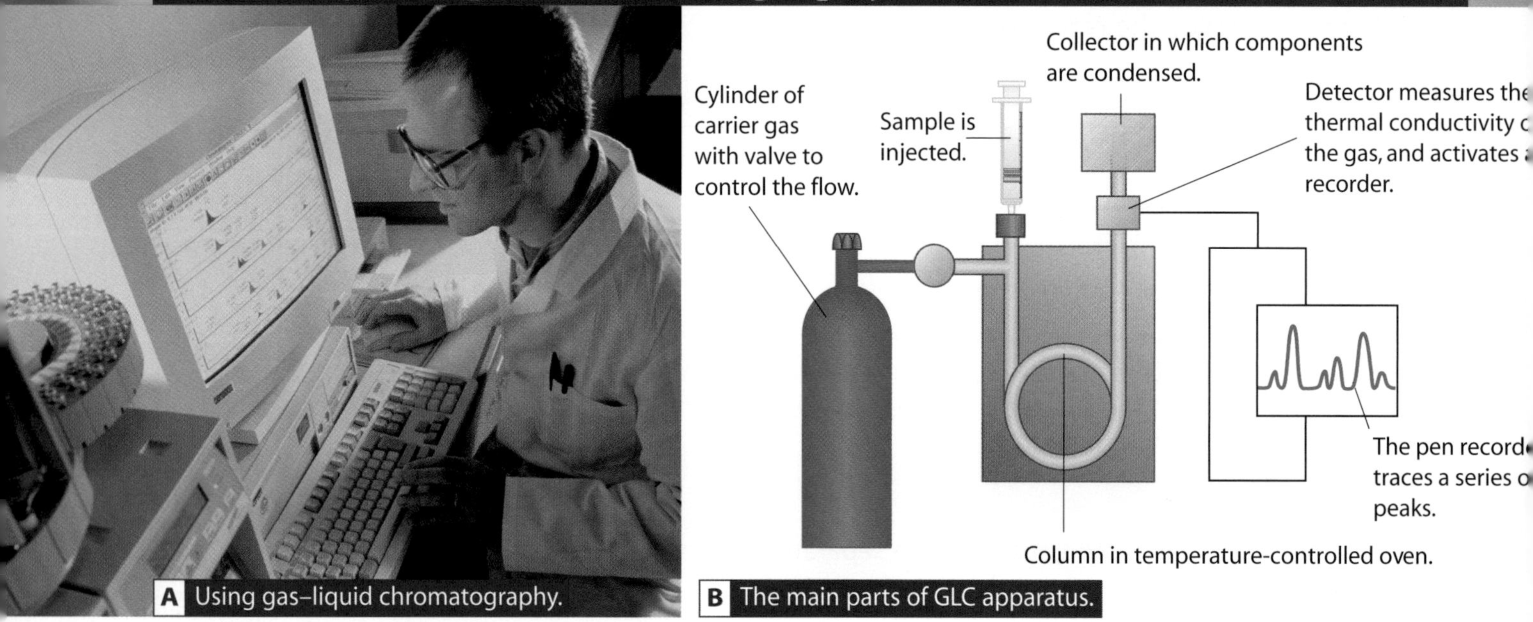

A Using gas–liquid chromatography.

B The main parts of GLC apparatus.

Gas–liquid chromatography (GLC) is used to separate and identify the components in tiny samples (0.000 001 g) of gases and liquids. It can also be used for **volatile** solids, which easily vaporise. It is also very important for finding amounts of each substance as it can provide quantitative results.

In gas–liquid chromatography the stationary phase is a solid packed into a glass column. The column may be very long (5–10 metres) and narrow (2–10 mm bore). To save space, the column is wound into a coil. A sample of a gas, liquid or volatile solid is injected into the apparatus and carried into the heated column by an unreactive gas, usually nitrogen or helium. As the components separate they are detected electronically.

?

1 List three differences between thin-layer chromatography and gas–liquid chromatography.

2 What is the mobile phase in gas–liquid chromatography?

Gas–liquid chromatography can be used to detect pollutants in the environment, identify samples of inks found on documents, detect traces of explosives found at crime scenes, reveal banned substances being used by athletes, or determine alcohol levels in blood samples from motorists.

During gas–liquid chromatography each substance in the mixture takes a different time, called the **retention time**, to pass through the column into the detector. Different substances are identified by comparing their retention times with those in tables of relative retention times.

C Gas–liquid chromatography is used to detect petrol or other chemicals used to start fires at a suspected arson site.

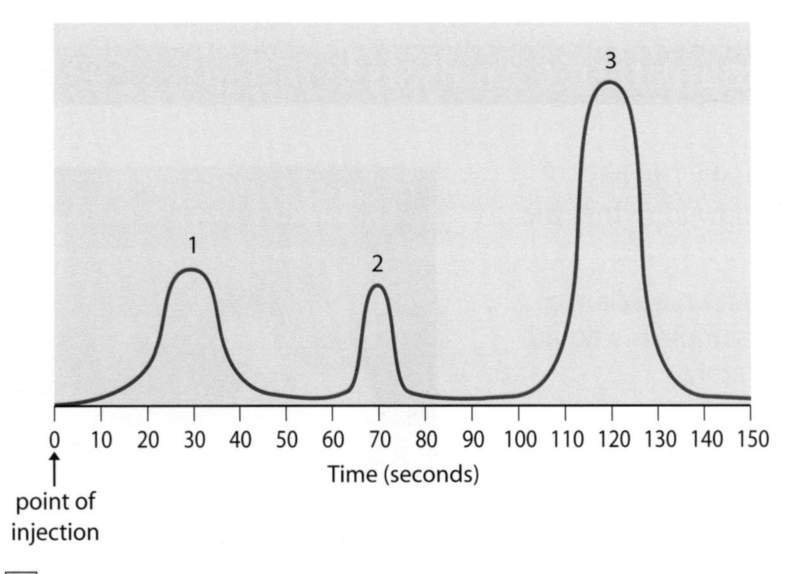

D

Diagram D shows a simple gas chromatogram. The distances between the point of injection and points 1, 2 and 3 represent the retention times of the three components of a mixture. The larger the area under each peak, the greater the percentage of that component in the mixture.

Columns may vary slightly in length, the bore of the tube, the pressure of the inert gas, or how tightly packed the stationary phase is. These will all affect the speed of flow through the column. Retention times are therefore compared to the time for the mobile phase to pass through the column, so that consistent standard data can be obtained for different substances.

?

3. Give three uses of gas–liquid chromatography.
4. How does the identification of substances using gas–liquid chromatography differ from that using paper and thin-layer chromatography?
5. Why would the length of the tube affect the retention time?
6. **a** Measure the retention time for each of the peaks in diagram D.
   b Which component is present in the largest amount?
7. Table E shows the retention times of some substances that may be present in the gas chromatogram shown in diagram D. Which of these substances were present?
8. How would you measure the actual percentages of each amount using a gas chromatogram?

Substance	Retention time (min)
A	16
B	30
C	70
D	93
E	118

E

Summary

Draw up a table to summarise the key features of gas–liquid chromatography, and compare it with thin-layer chromatography.

Quantitative analysis

What are the stages in a quantitative analysis?

Quantitative analysis must always be accurate. There is therefore a series of steps that must be followed so that the results are always reliable.

If the sample is a solid, it is first weighed out on a balance with an accuracy of at least 0.001 g. If the sample is a liquid the volume used must be measured accurately.

A Accuracy is essential when analysing the contents of medicines or baby foods.

?

1 Look at photo B. Which of these is the most accurate method of measuring 10 cm^3 of a liquid?

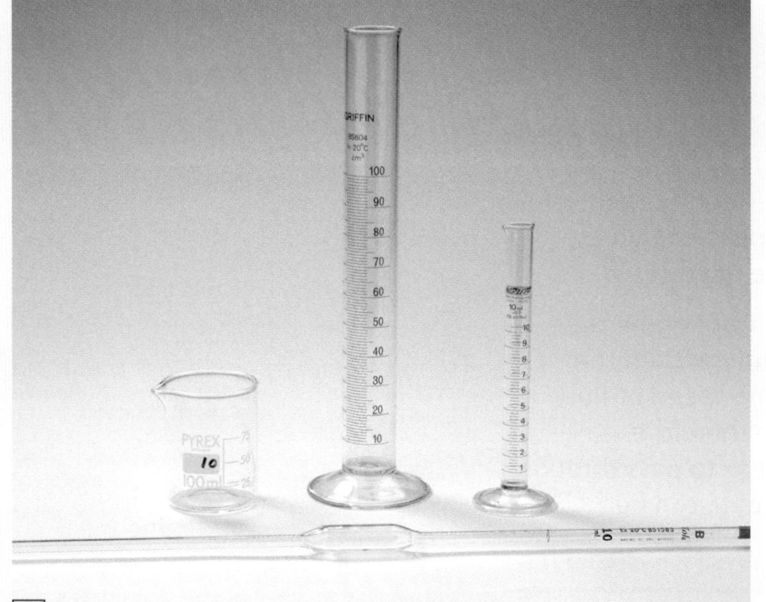

B

2 Would kitchen scales be suitable for carrying out an accurate analysis? Explain your answer.

The analyst must also work with replicate (repeat) samples, so that the results represent the bulk of the material being examined.

Once the sample has been accurately measured out, it is dissolved in water or another solvent to make up an exact volume of solution. This process is called **dissolving quantitatively**. In schools you use **standard flasks** to do this. These are manufactured so that they accurately contain 100 cm^3 or 250 cm^3 of a liquid.

C Standard flasks.

?

3 Why are accurate solutions made up in standard flasks rather than beakers?

A specific property of the solution must be measured quantitatively. This is often done using a titration, or may depend on another property such as colour. For example, copper sulfate solutions are blue.

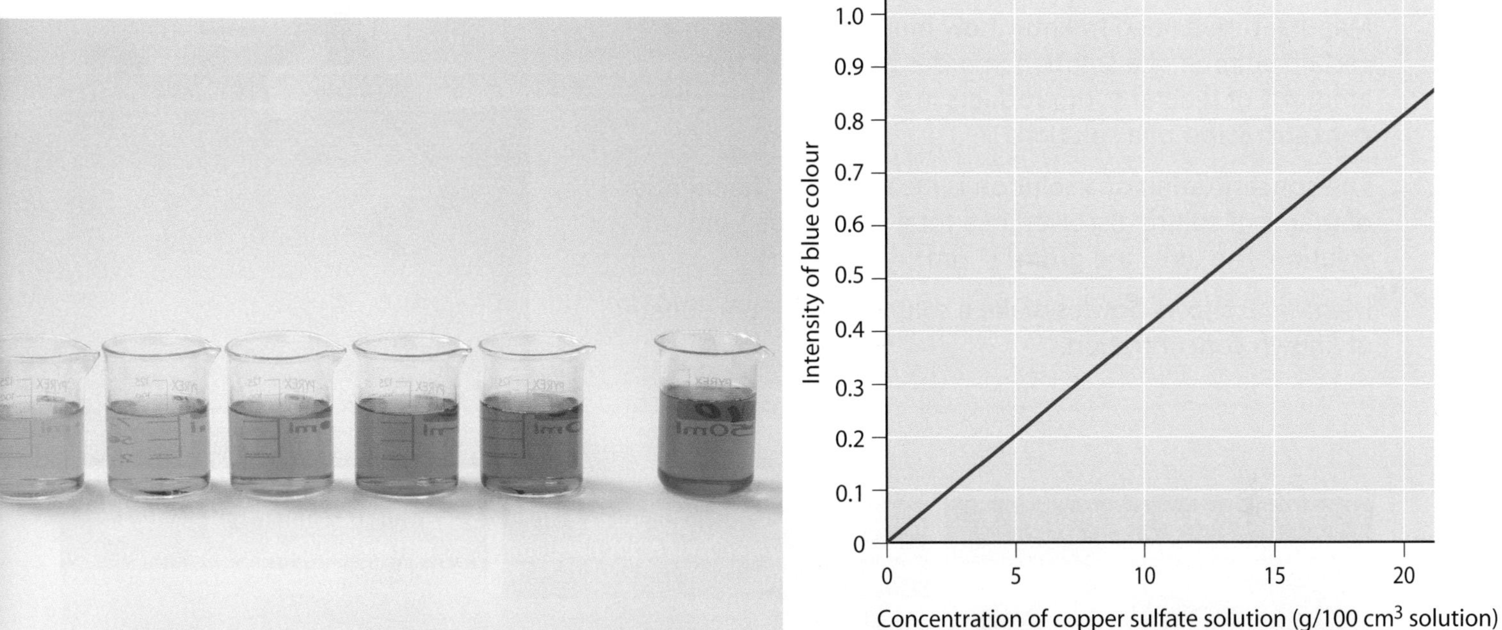

D Copper sulfate solutions of different concentrations.

E A calibration curve for copper sulfate solution.

To find the concentration of a solution of copper sulfate, a series of solutions containing different known concentrations of copper sulfate in water are prepared. The intensity of the colour of each can be measured using a colorimeter, and the results used to draw a **calibration curve** (diagram E). The intensity of colour in the solution of copper sulfate being investigated is then measured and the concentration is read off the curve.

In this example a calibration curve has been used to calculate a value for concentration. Titrations can also be used to find the concentration of a solution.

The degree of uncertainty in the results must also be estimated. Uncertainty is reduced by repeating the experiment until the results are consistent. The mean is then used as the best estimate of the true value for the result.

?

4 The intensity of the copper sulfate solution Q is 0.6. What is the concentration of this solution?

5 Another sample showed an intensity of 0.92. Can the concentration of this solution be obtained from the calibration curve? Explain your answer.

?

6 A titration gives these results for the volume of hydrochloric acid reacting with 10 cm^3 of potassium hydroxide solution: 12.3 cm^3, 15.6 cm^3, 12.5 cm^3, 12.2 cm^3.

a Which result should be omitted when calculating the mean? Explain your answer.

b The acid used was found to be more concentrated than had been stated. How would this affect the calculated concentration of the potassium hydroxide?

Summary

Draw a flow chart showing the main steps in carrying out quantitative analysis.

C7.14

Solution concentrations

How do we work out the concentration of solutions?

Manufacturers need to know how much chemical is in a certain volume of a solution in order to work out the amounts of reactants or products in a reaction. This is the **concentration** of a solution.

The concentration of a solution is measured as the number of grams of solute dissolved in a total volume of 1 dm^3 of solution. The units are g/dm^3 (1 dm^3 = 1000 cm^3).

Diagram B shows how to make a solution of copper sulfate of known concentration.

A Correct concentrations of copper sulfate must be used when spraying pests on grape vines.

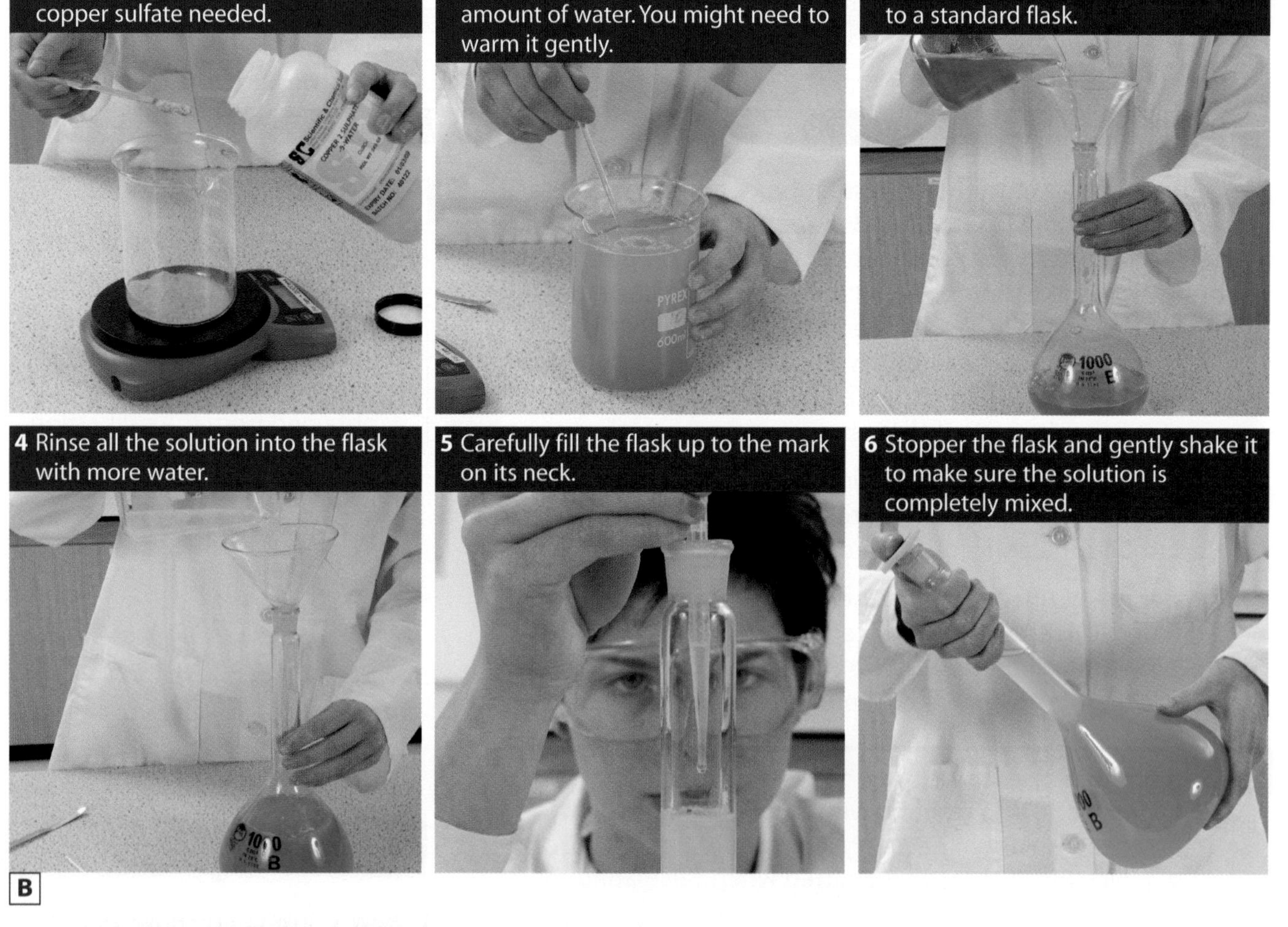

1 Carefully measure out the mass of copper sulfate needed.

2 Dissolve the copper sulfate in a small amount of water. You might need to warm it gently.

3 Transfer the contents of the beaker to a standard flask.

4 Rinse all the solution into the flask with more water.

5 Carefully fill the flask up to the mark on its neck.

6 Stopper the flask and gently shake it to make sure the solution is completely mixed.

B

?

1 Why might you need to warm gently in step 2?

2 Why is all the solution rinsed into the flask in step 4?

H Calculating concentrations

E You can calculate the concentration if you know the mass of the solute and the volume of the solution:

$$\text{concentration (g/dm}^3\text{)} = \frac{\text{mass of solute (g)}}{\text{volume (dm}^3\text{)}}$$

It is often more convenient to measure volumes in cm^3, so we use this equation:

$$\text{concentration} = \frac{\text{mass of solute (g)} \times 1000}{\text{volume (cm}^3\text{)}}$$

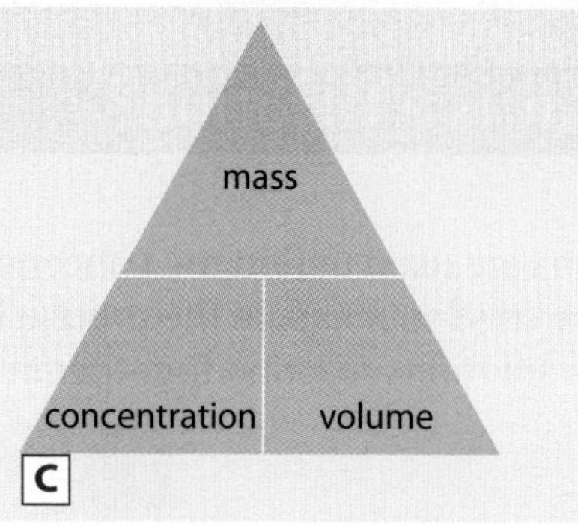

C

200 cm^3 of solution contains 10 g of sodium hydroxide. What is its concentration?

$$\text{concentration} = \frac{10 \times 1000}{200} = 50 \text{ g/dm}^3$$

E If we know the concentration and volume of the solution, we can calculate the mass of solute using the equation:

$$\text{mass of solute (g)} = \text{concentration (g/dm}^3\text{)} \times \text{volume (dm}^3\text{)}$$

$$\text{or} = \text{concentration (g/dm}^3\text{)} \times \frac{\text{volume (cm}^3\text{)}}{1000}$$

A solution of calcium chloride has a concentration of 24 g/dm^3. How much calcium chloride is there in 500 cm^3 of solution?

$$\text{mass} = \frac{24 \times 500}{1000} = 12 \text{ g}$$

?

3 Copy and complete this table.

Compound	Mass (grams)	Volume (cm^3)	Concentration (g/dm^3)
Potassium iodide	3	120	
Sugar	25	250	
Calcium chloride		100	30
Copper sulfate		250	8

4 A student made two different solutions of sodium chloride.
Solution A = 2 g dissolved in 200 cm^3 water
Solution B = 4 g dissolved in 500 cm^3 water
a Calculate the concentrations of these solutions.
b Which solution had the greater concentration?

5 20 g of potassium nitrate dissolves in 100 g of water at 20 °C, and 150 g dissolves in 100 g of water at 80 °C. What does this tell us about its solubility?

?

6 The maximum amount of fluoride that should be added to drinking water is 1 ppm (1 g of fluoride in 1 million grams of water). Is water with a concentration of 0.02 g in 10 000 grams of water safe? Explain your answer.

Summary

Write a set of instructions for making up a solution of iron sulfate of concentration 25 g/dm^3.

C7.15

Titrations

What is a titration?

Titrations are used to find the concentrations of solutions. They are used to measure the precise volumes of acidic and alkaline solutions reacting together in neutralisation reactions.

A Vinegars contain ethanoic acid.

To find the concentration of ethanoic acid in vinegar, we use an alkali of known concentration. The steps in carrying out this titration are as follows:

1. Use a **pipette** to place a measured volume of vinegar into a conical flask.
2. Add an **indicator** and place the flask on a white tile. Methyl orange is red in acid and yellow in alkali. A colour change occurs at the **end-point** where the reaction is complete.
3. Fill the **burette** with alkali.
4. Write down the reading on the burette, and then open the tap to let out 1 cm^3 of the alkali into the flask.
5. Swirl the flask gently.
6. Add another 1 cm^3 of the alkali from the burette and swirl the mixture again. Continue until the indicator changes colour.
7. Write down the volume of alkali added at the end-point.
8. Wash out the conical flask and repeat steps 1–7, this time adding the alkali drop by drop near the end-point. This will give a more accurate result.
9. Repeat the whole titration until you have two readings within 0.1 cm^3 of each other. This makes the results more reliable.

B

?

1. What will the colour change be at the end-point?
2. Why is this method unsuitable for brown malt vinegar?
3. Why are repeat readings within 0.1 cm^3 obtained in step 9?
4. Suggest some sources of error when you carry out a titration.

E We can use the results to calculate the concentration of the acid, using this equation:

$$m_2 = \frac{m_1}{M_1} \times \frac{V_1}{V_2} \times M_2$$

This gives the mass of solute in 1 dm^3 of solution, where:

V_1 = volume in cm^3 of solution 1 used
M_1 = relative formula mass of the solute in solution 1
m_1 = mass of solute in 1 dm^3 of solution 1
V_2 = volume in cm^3 of solution 2 used
M_2 = relative formula mass of the solute in solution 2
m_2 = mass of solute in 1 dm^3 of solution 2.

10 cm^3 of vinegar was neutralised by 20 cm^3 of sodium hydroxide of concentration 20 g/dm^3. What is the concentration of the ethanoic acid in the vinegar?

Solution 1 is the sodium hydroxide.

Therefore $V_1 = 20$, $m_1 = 20$, $M_1 = 40$.

Solution 2 is the ethanoic acid solution, so $V_2 = 10$, $M_2 = 60$.

Therefore $m_2 = \frac{20}{40} \times \frac{20}{10} \times 60 = 60$ g per dm^3 of solution.

?

5 25 cm^3 of sodium hydroxide solution (solution 2) was neutralised by 20 cm^3 of hydrochloric acid of concentration 7.3 g/dm^3 (solution 1). What is the concentration of the sodium hydroxide?

6 10 cm^3 of nitric acid was neutralised by 15 cm^3 of potassium hydroxide of concentration 2.8 g/dm^3. What is the concentration of the nitric acid?

C Kitchen descaler contains acid. How can the concentration of the acid be found?

Summary

Describe how to find the concentration of the acid in the kitchen descaler shown in photo C. Give reasons for as many of the steps as you can.

C7.16 More complex titrations

Why do we need RFMs in titration calculations?

H The equation that you used in Topic C7.15 works only for neutralisation reactions where the acid and alkali react together in a 1:1 ratio:

ethanoic acid + sodium hydroxide → sodium ethanoate + water

$CH_3COOH(aq) + NaOH(aq) \rightarrow CH_3COONa(aq) + H_2O(l)$

However, many neutralisation reactions have reactants that do not react in a 1:1 ratio. For example, 1 molecule of sulfuric acid reacts with 2 molecules of potassium hydroxide:

$H_2SO_4(aq) + 2KOH(aq) \rightarrow K_2SO_4(aq) + 2H_2O(l)$

We therefore need to adapt the calculation used in Topic C7.15 to take this into account. We do this by using relative formula masses and concentrations (see Topic C7.14). In the example below we have a 1:1 ratio of acid to alkali.

10 cm^3 of sodium hydroxide was neutralised by 15 cm^3 of nitric acid of concentration 6.3 g/dm^3. Calculate the concentration of the sodium hydroxide.

First write the balanced symbol equation for the reaction, with relative formula masses (RFMs) in grams written below the formulae:

$HNO_3(aq)$ +	$NaOH(aq)$ →	$NaNO_3(aq)$ +	$H_2O(l)$
63 g	40 g	85 g	18 g

Then find the mass of acid used:

15 cm^3 of nitric acid of concentration 6.3 g/dm^3 contains $\frac{15 \times 6.3}{1000} = 0.0945$ g acid.

63 g of nitric acid reacts with 40 g of sodium hydroxide.

Therefore 0.0945 g nitric acid reacts with $\frac{0.0945}{63} \times 40 = 0.06$ g alkali.

So 10 cm^3 of alkali contains 0.06 g sodium hydroxide.

1000 cm^3 (1 dm^3) contains 6.0 g sodium hydroxide.

?

1 Check that this is the correct answer using the method of Topic C7.15.

The same method can be used if the reactants are not in a 1:1 ratio.

A

Washing soda contains sodium carbonate, Na_2CO_3.
A solution of this can be analysed by titrating it against hydrochloric acid of a known concentration.

10 cm^3 of sodium carbonate solution was neutralised by 25 cm^3 of hydrochloric acid of concentration 7.3 g/dm^3:

$$Na_2CO_3(aq) + 2HCl(aq) \rightarrow 2NaCl(aq) + H_2O(l) + CO_2(g)$$

106 g 73 g 117 g 18 g 44 g

25 cm^3 of acid of concentration 7.3 g/dm^3 contains $\frac{25 \times 7.3}{1000} = 0.1825$ g HCl.

73 g of hydrochloric acid reacts with 106 g of sodium carbonate.

Therefore 0.1825 g hydrochloric acid reacts with $\frac{0.1825}{73} \times 106 = 0.265$ g Na_2CO_3.

Therefore 10 cm^3 of sodium carbonate contains 0.265 g.

So 1000 cm^3 (1 dm^3) of sodium carbonate contains $\frac{0.265 \times 1000}{10} = 26.5$ g.

?

2 Calculate the concentration of potassium hydroxide solution if 25 cm^3 of this reacts exactly with 50 cm^3 of sulfuric acid of concentration 9.8 g/dm^3.

3 25 cm^3 of sodium hydroxide solution of unknown concentration was titrated with dilute sulfuric acid of concentration 4.9 g/dm^3. 20.0 cm^3 of the acid was needed to neutralise the alkali. Find the concentration of the sodium hydroxide solution.

4 Show that you cannot use the method from Topic C7.15 for this reaction.

Summary

Write a list of the steps you need to take to calculate an unknown concentration from the results of a titration.

C7.17

Bulk and fine chemicals

What is the difference between bulk and fine chemicals?

The chemical industry makes some chemicals on a very large scale. Examples of **bulk chemicals** are ammonia, sulfuric acid, sodium hydroxide and phosphoric acid.

A This photo shows electrolysis. Sodium hydroxide is made by the electrolysis of brine (saturated salt solution).

Bulk chemical	Annual world production*	Materials needed	Main uses of the chemical
Ammonia NH_3	140 million tonnes	nitrogen from air, hydrogen from natural gas	80% fertilisers 5% nitric acid 15% other uses
Sulfuric acid H_2SO_4	120 million tonnes	sulfur from the ground or natural gas, oxygen from air, water	31% fertilisers 18% paints and pigments 12% fibres
Sodium hydroxide NaOH	41 million tonnes	rock salt from the ground	34% various uses, e.g. soaps and foods 30% manufacture of speciality chemicals 20% other chemicals 16% fibres
Phosphoric acid H_3PO_4	48 million tonnes	phosphate rock, sulfuric acid	80% fertilisers 12% detergents 8% animal and human food

*Data from *The Essential Chemical Industry* (RSC, 1999).

B

?

1. Which bulk chemical is made in the greatest amount?
2. What raw materials are used to make sulfuric acid?
3. Which bulk chemicals are important in farming? Explain your answer.

Fine chemicals (or **speciality chemicals**) are chemicals that have to be very pure. They are made in much smaller amounts than bulk chemicals, but have much higher prices because they cost more to manufacture. Fine chemicals include drugs, food additives and fragrances.

Drugs are substances that affect chemical reactions in your body. They include commonly used substances such as alcohol and nicotine, illegal substances such as cocaine, and thousands of medicines made by the pharmaceutical industry.

Medicines must be effective (must work) if they are to sell. They also need to be safe, have few side-effects and be easy to take. They should also be made from chemicals that are easily obtainable and will not run out.

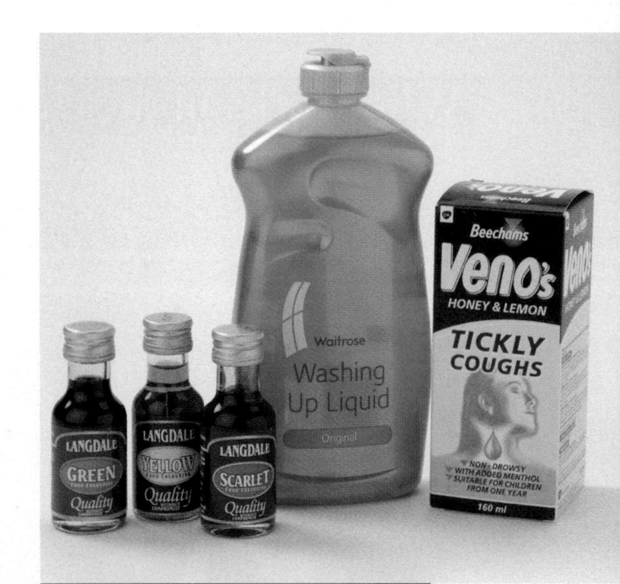

C These all contain fine chemicals.

? 4 Name and explain three important properties that a medicine should have.

Food additives may be added to food during its production. The main types include anti-oxidants, colours, emulsifiers, flavour enhancers, preservatives and sweeteners. There are at least 60 different anti-oxidants and about the same number of colours. Each additive needs a different chemical process to make it.

? 5 Write down three foods that have had colour added to them.

6 Why are so many sweeteners used?

D All these perfumes contain different mixes of chemicals so that they smell different from one another.

A particular **fragrance** comes from a mix of different chemicals. The smell of bitter almonds contains only about six different chemicals, but the smell of strawberries contains more than 1500 different chemicals. In the perfume industry, these chemicals are made in tiny amounts and then mixed in very carefully calculated quantities to make the right smell.

? 7 State two things other than smell that should be considered when making a perfume from chemicals.

8 Which would be more profitable – bulk or speciality chemicals? Explain your answer.

Summary

Write encyclopaedia entries for 'bulk chemicals' and 'fine chemicals', including examples of each.

C7.18

Developing new chemicals

Who works in the chemical industry?

I'm Henry Davies and I'm the Chief Executive of this factory where we make cough medicines. This involves using fine chemicals. Many businesses have switched to making fine chemicals, as we can make more profit. This is because other countries started to make bulk chemicals at a cheaper price and the cost of producing these chemicals in this country increased too much.

A

Hi, I'm Andy. It's my job to try to make new chemicals by starting with known compounds and changing them chemically so that a new drug is formed. It's hard work and can be quite tedious, but when we find something that might work well it's really exciting. At the moment I'm looking for a new catalyst to speed up one of the reactions.

B

My name is Sally and I work in the research and development department of this company. We spend a lot of money on research, in the hope of finding a really effective medicine that doesn't have any side-effects. I then help to develop any new medicines that may look hopeful and oversee all of the safety tests that must be carried out. The medicines are also tested for purity by the quality control department.

C

My name is Alex and I look after the finances of the company, so that we make a profit. The money we get from selling the medicines must be enough to pay for the raw materials, the energy used, wages, designing and building the plant, waste disposal, marketing and sales, research and development, and the repayment of any loans that were needed to set up the company.

D

I'm Becky and I'm in charge of health and safety here. The government has strict rules and regulations to control the chemical processes that we carry out, as well as the way in which we store and transport our chemicals. The process engineers keep the equipment running correctly, but there is always a risk of an accident. We try to protect the people who work here as well as other people and the environment. We also employ an environmental officer.

E

I'm Bill and I deliver oil, which is a raw material that this company uses. The safety symbols on my tanker identify the hazardous substance that I carry, and also the type of hazard. In case there is an accident, they also tell the fire service what substances to use to neutralise the risk and what type of breathing apparatus and protective clothing to wear.

F

?

1. Why do cough medicines need fine chemicals?
2. Why must drugs be tested for safety?
3. What would be the advantage of finding a catalyst that speeds up a reaction?
4. Why is research and development so expensive?
5. Give two examples of how chemicals could damage the environment.
6. What is the main hazard in transporting oil products?
7. What would the company have to do if oil got too expensive to use as its raw material?

Summary

Make a poster advertising some of the job vacancies for specialist chemists in a chemical company.

C7.19 Producing chemicals

What must be considered when we make new chemicals?

Making new chemical molecules from simple starting materials is called **synthesis**. Synthesis may involve several stages, each of which has its own specific reaction conditions.

Chemical processes start with **feedstocks**. These may be obtained from natural resources, but there is concern that some of these are finite and may run out. Feedstocks may also be chemicals already produced by the chemical industry. For example, fertilisers are made from ammonia produced by the Haber process.

A Rock salt (above) and the crude oil that is processed at this refinery are important raw materials for the chemical industry.

? 1 Name two finite resources.

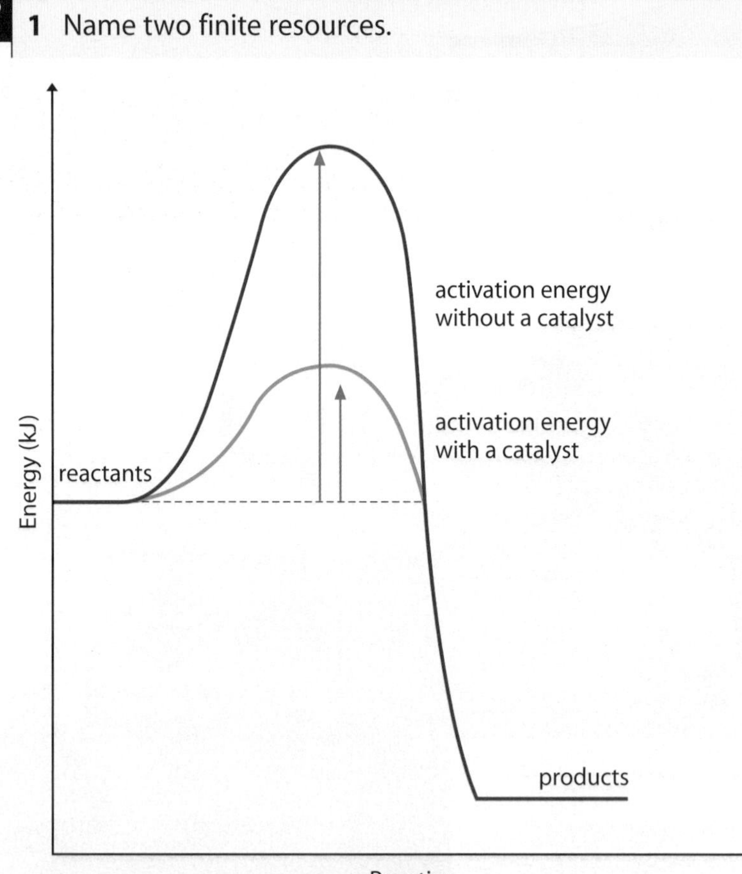

B A graph to show the effect of a catalyst on activation energy.

A synthesis reaction often involves reacting chemicals together under conditions of high temperature and pressure. This may be expensive, so catalysts are often used. Each reaction has an activation energy, which is the minimum amount of energy needed to get the reaction going. Catalysts produce an alternative route for the reaction with lower activation energy, so the reaction goes faster. The temperature needed for the reaction may then be lowered, saving energy.

? 2 **a** How do catalysts work?
b Why are they used in chemical reactions?

Once the reaction has taken place, the products must be separated. Many reactions, especially those involving organic chemicals, do not give very high yields – their **atom economy** is low. This is wasteful, as it may not be possible to reuse unused starting material. Other **by-products** may also be formed and these have to be separated from the desired product.

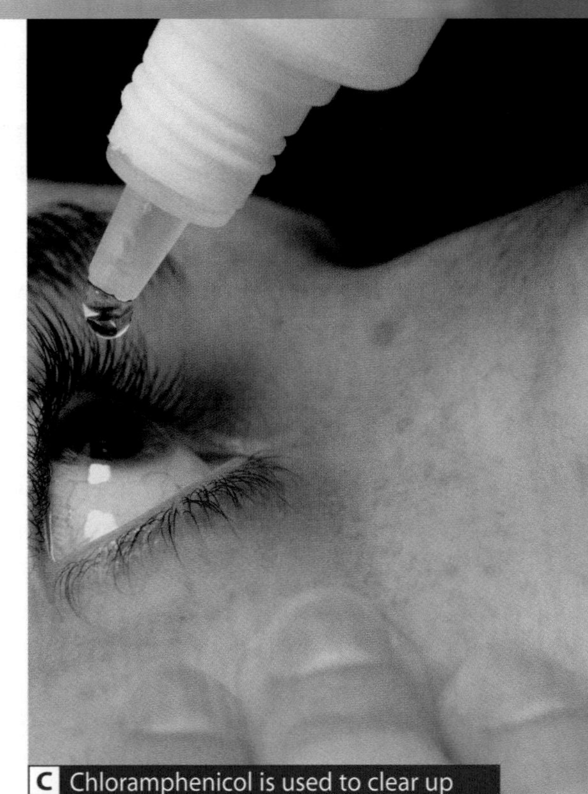

C Chloramphenicol is used to clear up conjunctivitis.

?

3 a The chloramphenicol in the eye drops in photo C can be made by two different methods. One has an atom economy of 74% and the other of 63%. Which reaction would be better to use?

b Suggest a reason why the atom economies are so low.

Many by-products have no use, and some may be dangerous to store. These must be stored and disposed of safely. However, some by-products can be useful. For example, when fats react with sodium hydroxide to form soaps, glycerol is also made. This can be used to make explosives and printing ink, and is also used in foods.

?

4 Suggest how toxic by-products could be stored.

Once the product has been purified, its purity must be checked using analytical techniques such as chromatography and titration.

The purity of chemicals used in industry depends on the final use of the chemical. Many bulk chemicals are sold to other manufacturers without purification. This is because they will be converted into other chemicals, perhaps in a series of steps where yields are low and impurities are introduced. Purification is then carried out at the end of the manufacturing process. Fine chemicals are much purer, as they have specific uses where purity is very important. Impure phosphoric acid is converted directly into fertiliser, while much purer phosphoric acid is needed in the food industry.

?

5 If a product produces three spots during chromatography, is it pure or not? Explain your answer.

6 How could you use titration to check whether phosphoric acid is pure?

Summary

Draw a flow diagram showing what happens during a chemical synthesis.

C7.20

Green chemistry

What are the characteristics of green chemistry?

Many people think of the chemical industry as being like photo A, or releasing toxic chemicals into rivers and killing fish. While this may have been true of a few companies in the past, there is now a move to 'clean up the image' of industry. This involves using the principles of **green chemistry**, which also lead to the sustainability of chemical processes.

A Some parts of the chemical industry as they used to be.

There are 12 principles of green chemistry:

1. *Reduce pollution of the environment* by designing chemical processes that produce little or no waste to clean up.
2. *Produce safe, non-toxic and effective chemicals* by designing new ones if necessary.
3. *Improve safety* by designing chemical syntheses that do not produce toxic products or by-products. Making a specific chemical may include several steps, each of which should be as safe as possible.
4. *Avoid depleting finite resources* like oil and coal by using renewable raw materials and feedstocks. Renewable feedstocks are often made from wastes from other processes.
5. *Reduce waste* by using reusable catalysts that reduce the amount of chemicals needed.
6. *Avoid unnecessary chemical reactions* that use extra chemicals and produce extra waste.
7. *Maximise the atom economy*. The higher the percentage yield, the fewer reactants are needed and the fewer chemicals are wasted.
8. *Use safer solvents and reaction conditions*. In the past solvents such as CFCs were used. However, since their impact on the ozone layer has been recognised, their use has been reduced and their manufacture banned.
9. *Save energy* by lowering the temperature and pressure of any chemical reaction.
10. *Design biodegradable products* that are broken down naturally. For example, many plastics are not biodegradable and can stay in landfill sites for hundreds of years. This could be avoided if more biodegradable plastics were used.
11. *Include real-time monitoring and control* to eliminate the production of any by-products.
12. *Minimise the chance of accidents*. All of the processes should be as safe as possible to prevent explosions or the release of toxic substances into the atmosphere.

B The chemical industry now.

?

1 What will happen to a process that depends on oil when the oil runs out?

2 How can acidic wastes be made harmless?

3 Why is the production of Marmite® from by-products of the brewing industry a good example of green chemistry?

4 Which of the principles of green chemistry depend on good design of the chemical processes and plant?

5 Why do catalysts reduce the amount of energy needed?

6 What is the advantage of making biodegradable products?

7 **a** Which would be better – a four-step process that was totally safe, or a two-step process that involved using cyanide?
b What other factors might need to be considered before making a final decision?

8 What are the social and economic benefits of green chemistry?

9 How is green chemistry different from environmental chemistry (the chemistry of the natural environment)?

Summary

Write an article for the school magazine about green chemistry.

Fermentation

How is fermentation used to make ethanol?

A Ethanol (available at petrol pumps like this one) is an important fuel, especially in South America.

Ethanol is an important chemical with a wide variety of uses. More than 330 000 tonnes of pure ethanol are produced in this country each year. It is used as a fuel, as a feedstock for other chemical processes such as the manufacture of fibres and esters, and as a solvent in the manufacture of cosmetics, pharmaceuticals, detergents and inks.

Ethanol has been known for centuries, as it is produced during the process of **fermentation**. Yeast is added to a sugar solution and left for several days in the absence of air. These **anaerobic** conditions cause zymase enzymes in the yeast to convert the glucose into ethanol and carbon dioxide:

$$C_6H_{12}O_6(aq) \rightarrow 2C_2H_5OH(aq) + 2CO_2(g)$$

The **optimum** temperature for fermentation is 37 °C, as yeast grows best at this temperature. A pH near 7 (neutral) also gives the yeast the best conditions for growth.

The starting sugars can be obtained from sugar cane, sugar beet or maize. Because so much sugar cane is grown in South America, especially Brazil, ethanol is a particularly important fuel for cars there.

Ethanol obtained by this process is no more than 15% pure. This is because greater concentrations of ethanol denature the enzymes in the yeast and the reaction stops.

?

1. Write a word equation showing what happens when ethanol burns as a fuel.
2. Ethanol has a lower boiling point than water. Why does this make it a better solvent than water for perfumes and deodorants?

Home beer and wine-making kits use the same method. Yeast is added to a sugar solution containing either a beer extract (made from hops and malted barley) or a wine extract (made from grapes). These are left for several weeks, so that air cannot enter.

?

3 What products could have been made by fermentation 5000 years ago?

4 What happens when an enzyme is denatured?

5 Why is the wine kit shown in photo B left in an airing cupboard?

6 How could you try to find the effect of pH on the rate of fermentation?

B Making wine at home.

C Whisky is produced by the distillation of a more dilute ethanol solution.

The ethanol solution can be concentrated by fractional distillation to make products like whisky and brandy. Ethanol boils at 78 °C, so when an ethanol–water mixture is heated, the ethanol boils at a lower temperature and is separated. The liquid that distils over at 78 °C is 96% pure ethanol. It is impossible to remove the final 4% of water by distillation, so alternative ways are used to make pure ethanol.

?

7 How could you show that ethanol is useful as a fuel?

8 a Explain how the fermentation method of producing ethanol uses green chemistry.
  b In what ways is the process not green?

Summary

Make a leaflet for a whisky manufacturer to give to customers, explaining how whisky is made.

C7.22

New ways of making ethanol

How is ethanol made without using fermentation?

A The hydration of ethene to produce pure ethanol.

Although fermentation has been used for thousands of years, it is slow and produces only a 15% solution of ethanol. There are two more modern methods of producing ethanol.

Pure ethanol can be produced by reacting ethene with steam, a process called **hydration**:

ethene + steam → ethanol

$C_2H_4(g) + H_2O(g) \rightarrow C_2H_5OH(g)$

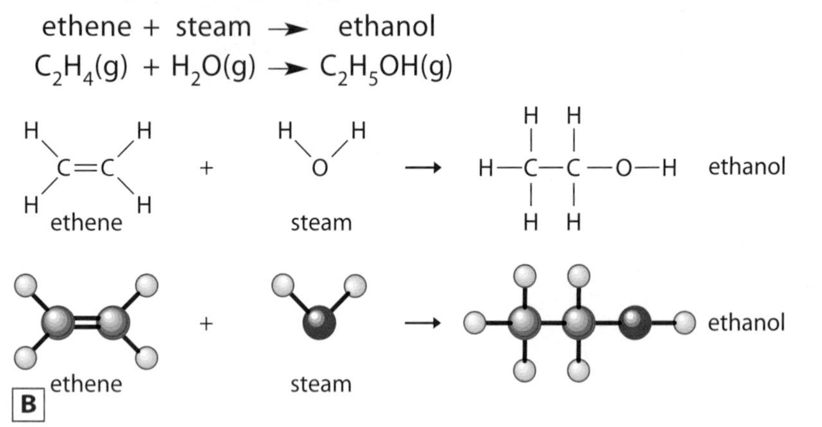

B

The ethene is obtained from the refining of oil, and hence comes from a finite resource. Ethene and steam are mixed together, and then passed continuously over a catalyst of phosphoric acid. The temperature needed by this reaction is 300 °C at a pressure of 60–70 atmospheres.

?

1. Why is fermentation slow?
2. Why is the ethanol produced in the hydration process a gas?
3. Is this process based on any principles of green chemistry? Explain your answer.

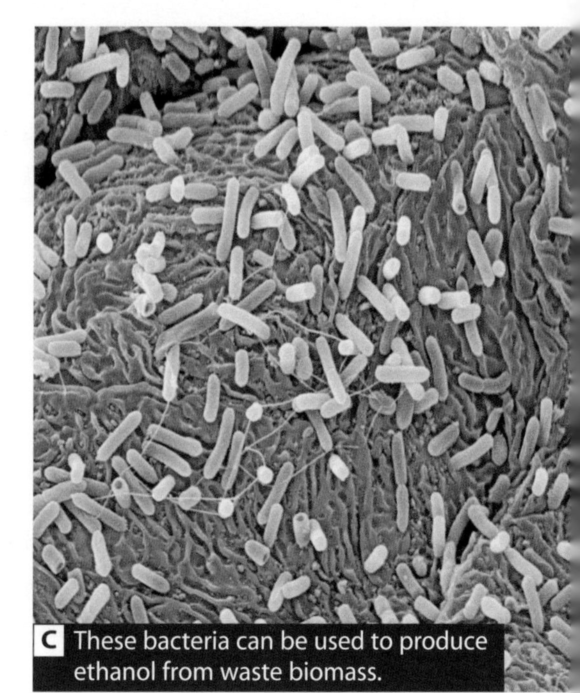

C These bacteria can be used to produce ethanol from waste biomass.

An alternative and very new method of making ethanol involves the use of genetically modified *E. coli* bacteria. These bacteria have had a gene introduced into them that is capable of converting all types of carbohydrates found in plant cells into ethanol. Until now, yeast fermentation has been carried out on valuable food materials such as cornstarch and sugar from sugar cane. Their use as a feedstock for fermentation meant that they could not be used to feed people and animals. However, the new genetically modified *E. coli* can produce ethanol from waste biomass such as sugar cane residues, corn husks, wood and other waste organic material.

The biomass is mixed with the bacteria at 30–37 °C and in slightly acidic conditions (pH 6.6). The ethanol produced is about 5% pure and can then be concentrated by fractional distillation.

?

4 What are the advantages of the *E. coli* method of producing ethanol?

5 What is the main disadvantage of this method?

Table D gives a comparison of the three methods of making ethanol.

	Fermentation	**Hydration of ethene**	**Using *E. coli* bacteria**
Resources used	renewable resources, e.g. sugar cane and maize	finite resources – ethene	renewable resources – waste biomass such as corn husks and wood
Type of process	batch process where everything is mixed together and left for several days; the ethanol is then removed and distilled	continuous process where the reactants are passed over the catalyst	batch process where the biomass and bacteria are mixed and left for several days; the ethanol formed must be distilled off
Reaction conditions	37 °C and atmospheric pressure	high temperature and pressure	30–37 °C and atmospheric pressure
Rate of reaction	slow	fast	slow
Purity of product	15% pure	96% pure	5% pure

D

?

6 Which method produces the purest ethanol?

7 Which method uses most energy?

8 Which method is most sustainable? Explain your answer.

9 How might the method of ethanol production be determined by where in the world you live?

Summary

Write a couple of paragraphs to compare and contrast the three methods of making ethanol. Explain which method is the most sustainable.

Revision questions

Mariam and Tom work for 'Fine Foods', where they make fine chemicals. They have been asked by a local food company to make some artificial pineapple flavour that could be added to foods. The compound that they want to make is ethyl butanoate, which is made from butanoic acid and ethanol.

1 a Draw the structural formula for:
  i ethanol
  ii butanoic acid (which contains four carbon atoms).

b What are the functional groups in:
  i an alcohol like ethanol
  ii a carboxylic acid like butanoic acid?

2 Write a word equation showing the reaction between ethanol and butanoic acid.

3 a What are fine chemicals?

b Is ethanol a bulk or a fine chemical? Explain your answer.

4 Give **two** uses for ethanol, other than making ethyl butanoate.

5 a The reaction between ethanol and butanoic acid is exothermic. Is more energy taken in when the bonds break or when the new bonds form?

b Draw a simple energy level diagram showing what happens in this reaction.

Tom was looking for a supply of pure ethanol. To obtain this, he could choose between either distilling some ethanol obtained by fermentation, or distilling some ethanol produced by the action of genetically engineered *E. coli* bacteria on waste biomass, or buying some pure ethanol made by the hydration of ethene. He knew that his demand for ethanol could vary according to how much of the pineapple flavour he could sell, and he also knew that he needed pure ethanol.

6 a Which of these three methods produces the purest ethanol?

b Which of these methods is the cheapest?

c Which method(s) use sustainable resources?

d What is the advantage of using waste biomass?

e Which of these methods could be regulated to meet sudden increases in demand for ethanol?

f Which of these methods uses green chemistry?

The reaction between ethanol and butanoic acid is in equilibrium.

7 **a** What does a state of equilibrium mean?
b How do we show an equilibrium reaction when we write an equation?

Mariam wanted to check the purity of the butanoic acid before she started the reaction. She did this by reacting the butanoic acid with a solution of potassium hydroxide, which was made up especially for this reaction.

8 Describe how to make a solution of potassium hydroxide of concentration 112 g/dm^3.

9 What is the name of the procedure that Mariam would use to check the concentration of the butanoic acid?

10 Butanoic acid and potassium hydroxide react together in a 1:1 ratio. What is the concentration of the butanoic acid if 100 cm^3 of acid reacts with 50 cm^3 of the potassium hydroxide solution?

The relative formula mass of butanoic acid is 88 and that of potassium hydroxide is 56.

Use the equation:

$$m_2 = \frac{m_1}{M_1} \times \frac{V_1}{V_2} \times M_2$$

where:

V_1 = volume in cm^3 of solution 1 used
M_1 = relative formula mass of the solute in solution 1
m_1 = mass of solute in 1 dm^3 of solution 1
V_2 = volume in cm^3 of solution 2 used
M_2 = relative formula mass of the solute in solution 2
m_2 = mass of solute in 1 dm^3 of solution 2.

11 What type of chromatography is most suitable for testing the purity of the butanoic acid? Explain your answer.

12 **a** Draw the apparatus that Tom and Mariam would use to make the ethyl butanoate on a small scale.
b Explain why this apparatus is needed.

13 Calculate the mass of butanoic acid that would react with 92 g of ethanol.

H **14** Write a balanced equation to show what happens when sodium is added to the ethanol used.

Pre-release question

You will get a pre-release paper before the C7 exam. The text and questions here are to show you the kinds of questions you might have to answer, and to give you some hints about how to use the pre-release passage to help you to revise for the exam.

Hydrogenating vegetable oils

There are several different types of fat that are found naturally in our diets. Fats from animals, such as milk, are high in saturated fats. Most oils of plant origin contain some saturated fats, but only certain oils such as palm or coconut oil have a high percentage of saturated fats. Vegetable oils contain mainly monounsaturated or polyunsaturated fats.

Unsaturated fats have lower melting points than saturated fats, which is why vegetable oils are liquids whereas sources of animal fat, such as butter or lard, are usually solids. To raise their melting points, oils can be hydrogenated. They are heated with hydrogen, when some of the double bonds break and join up with hydrogen, so reducing the number of double bonds in the molecule. This process also twists some of the molecules, to produce '*trans*' fatty acids.

The diagrams show two fatty acids with the formula $C_{18}H_{34}O_2$, both with one carbon–carbon double bond. Oleic acid is a '*cis*' acid, with the two hydrogen atoms joined to the carbon atoms either side of the double bond on the same side. Elaidic acid has these two hydrogens on opposite sides of the chain.

Fats are widely sold for cooking, but are also used by food manufacturers in making cakes, pies and biscuits, and for frying chips and other foods. For many of these uses, a fat that is solid at room temperature is needed, so manufacturers use hydrogenated vegetable oils. These hydrogenated oils also keep for longer.

Both saturated fats and '*trans*' fats are bad for our health. These fats in the diet tend to increase the amount of cholesterol in the blood, which is linked with a higher risk of heart disease.

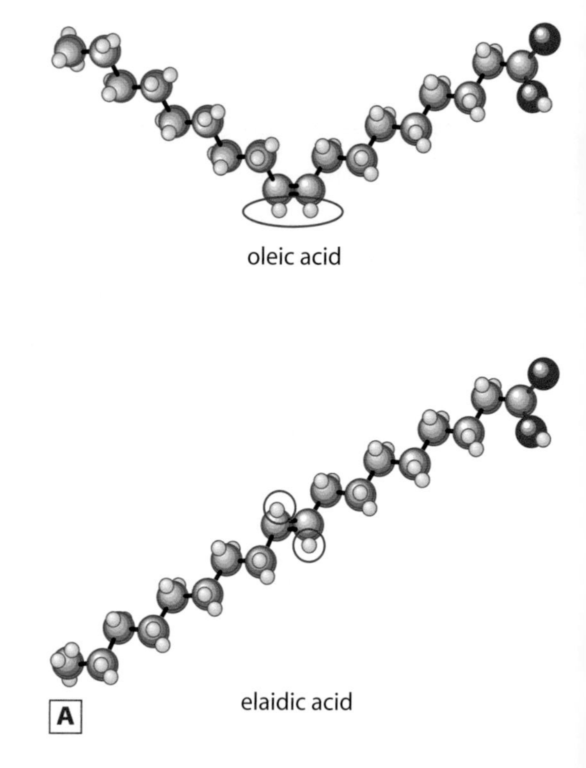

When you get your pre-release paper, read it carefully and then think about which areas of science it covers. Go back to the module where this was covered and revise any of the content that looks relevant to the passage. The modules you need may be in the GCSE Science or the GCSE Additional Science book. Also think about the 'Ideas about Science' that might be relevant to the text (see the list on page 63).

The passage above is about fatty acids, which are looked at in Module C7. It looks at the bonding between atoms in a molecule, which was covered in C5, and at the effects of diet on health, which were covered in C3.

Some of the 'Ideas about Science' that could be asked about include people's attitudes to risk, and the role of the government in regulating what goes into foods, both looked at in C3.

If the passage above was given to you as your pre-release paper, it would be a very good idea to revise what you learned in Modules C3 and C5. This doesn't mean that you *will* get questions on these modules, just that you *might*.

The questions below are similar to the type of question you may get in the exam. You will not see the questions before the exam, only the text.

1 **a** Explain the difference between a saturated and an unsaturated compound. [2 marks]

b What do you think 'monounsaturated' and 'polyunsaturated' mean? [2 marks]

2 **a** Why do manufacturers 'hydrogenate' some vegetable oils? [1 mark]

b Why could a saturated fat be referred to as 'natural' but a '*trans*' fat as 'artificial'? [1 mark]

3 **a** A consumer group states that 'hydrogenation is carried out only for the benefit of the food manufacturing companies'. Suggest a reason for this statement. [1 mark]

b Why might manufacturers claim that this is also in the interests of consumers? [1 mark]

4 Eating less saturated or '*trans*' fats could benefit our health.

a Why would we be healthier if the fats we ate were only monounsaturated or polyunsaturated fats? [1 mark]

b Many people know the risk of eating these kinds of fat. Suggest **two** reasons why they might prefer these fats. [2 marks]

5 Genetically modified foods have been on sale in the USA for many years, and no risks to health have been demonstrated. In the UK, most people will not buy GM foods, but are happy to continue eating saturated and '*trans*' fats.
Suggest why this is so, using ideas about perceived risk and actual risk.
One mark is for a clear, ordered answer. [3 marks + 1 mark]

GCSE Physics

GCSE Physics includes work from the GCSE Science book (Modules P1, P2 and P3) and from the GCSE Additional Science book (Modules P4, P5 and P6), as well as the work in this module.

The table shows how your GCSE Physics will be assessed. You may have already done the tests for Units 1 and 2, and possibly also some of the coursework.

Unit	Type of assessment	Tests you on ...
1	**Test paper** 40 minutes written paper (42 marks)	P1 The Earth in the Universe P2 Radiation and life P3 Radioactive materials
2	**Test paper** 40 minutes written paper (42 marks)	P4 Explaining motion P5 Electric circuits P6 The wave model of radiation
3	**Test paper with pre-release** 60 minutes written paper (55 marks)	P7 Further Physics Pre-release question (see below)
4	**Coursework** Practical Data Analysis (16 marks) Case Study (40 marks)	How well you can analyse and evaluate data from an experiment. See page 158. How well you can gather and interpret information on a scientific subject, and present conclusions. See page 156.
5	**Coursework** Practical Investigation (40 marks)	How well you can plan and carry out a full investigation, and how well you can interpret your data and evaluate your data and conclusions. See page 158.

Pre-release question

The pre-release question is a passage of text, with questions based on the text or on science related to the text. The science involved could be any topic from Modules P1 to P7. You will be given the passage of text (but not the questions) before the examination, so you have time to look up the science. The questions may also test what you know about 'Ideas about Science'. You have been taught about these 'ideas' throughout the modules (in Biology and Chemistry modules as well as Physics). There is a practice question of this kind on page 154.

Ideas about Science

The Ideas about Science are:

- **Data and their limitations** – why results can vary or be in error, and how to make the best estimate of a true value.
- **Correlation and cause** – how to investigate the relationship between a factor and the outcome of an investigation, the difference between correlation and cause, and the need for a plausible mechanism linking a factor and an outcome.
- **Developing explanations** – how scientists explain things we observe, and how their explanations are tested.
- **The scientific community** – how scientists work together, and how they check and accept (or reject) scientific findings and explanations.
- **Risk** – what risks are, how we perceive them, and why some risks are accepted when others are not.
- **Making decisions about science and technology** – how some aspects of science raise ethical issues, and how decisions about science depend on social and economic factors as well as scientific factors.

P7 Objectives

When you have studied Module P7 you should be able to:

- explain some of the apparent motions of the Sun, the Moon and the planets across the sky
- describe how astronomers give coordinates of stars and other objects in the sky, and why some stars can be seen only at certain times of the year
- describe how telescopes use lenses and mirrors, and calculate the angular magnification of a telescope
- explain what parallax is, and how astronomers can use it to calculate distances
- explain what a Cepheid variable is, and how such stars can be used to calculate distances
- recall the main issues in the debate between Curtis and Shapley, and how Hubble's observations helped to resolve the debate
- describe how stars are formed, and the changes that happen as they get older
- explain some of the stages in a star's life cycle in terms of the nuclear reactions taking place within the star
- describe suitable locations for telescopes, and the advantages and disadvantages of computer control and the use of remote telescopes
- understand why international collaboration is necessary in astronomy.

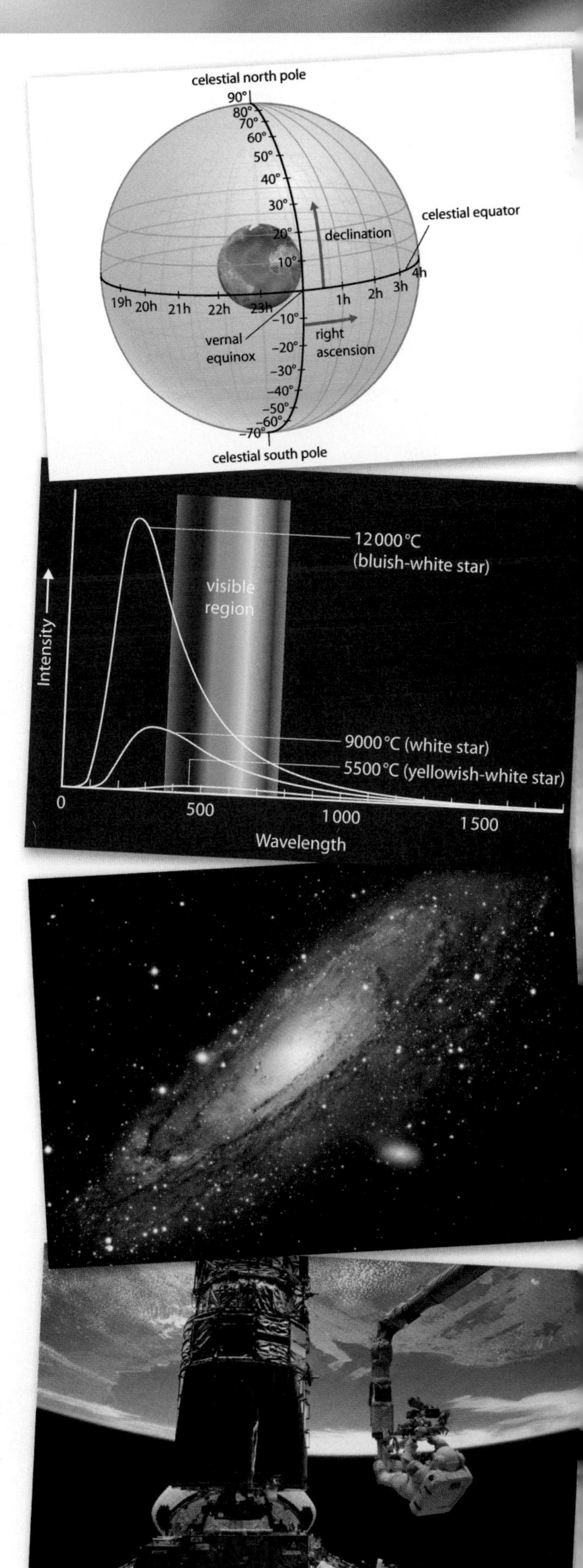

P7.1 Movements in the sky

Why do the Sun, Moon and stars appear to move?

A This photograph of the stars was exposed for 1 hour.

Studying the Universe is different from many other areas of science. Discoveries in science always involve observations, but in most areas of science scientists can set up experiments to test their ideas. In astronomy, we can only use observations.

The first astronomical observations were that the Sun appears to move across the sky from east to west once every 24 hours, and the Moon and stars also appear to move across the sky. These motions can be explained by the movements of the Earth and the Moon, as shown in diagram B.

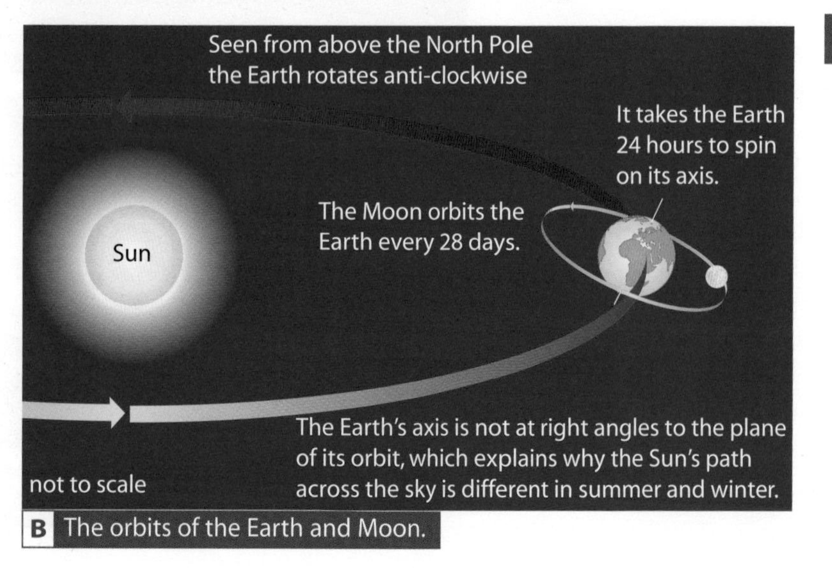

B The orbits of the Earth and Moon.

?

1 **a** How do the Sun, Moon and stars appear to move across the sky?
b Which part of the model shown in diagram B explains these movements?

2 **a** Describe the differences in the Sun's apparent movement between summer and winter.
b What is shown in diagram B that explains this?
c How does diagram B show why it is summer in New Zealand when it is winter in the UK?

A 'day' is usually described as being 24 hours long. This is a **solar day**, which is the time from the Sun being at its highest point in the sky to the next time at which it is at its highest. However, days can also be measured in relation to the stars, which are so far away that they appear to be fixed in the sky. One day measured using the stars is called a **sidereal day**. A sidereal day is slightly shorter than a solar day because the Earth is moving around its orbit at the same time as it is rotating.

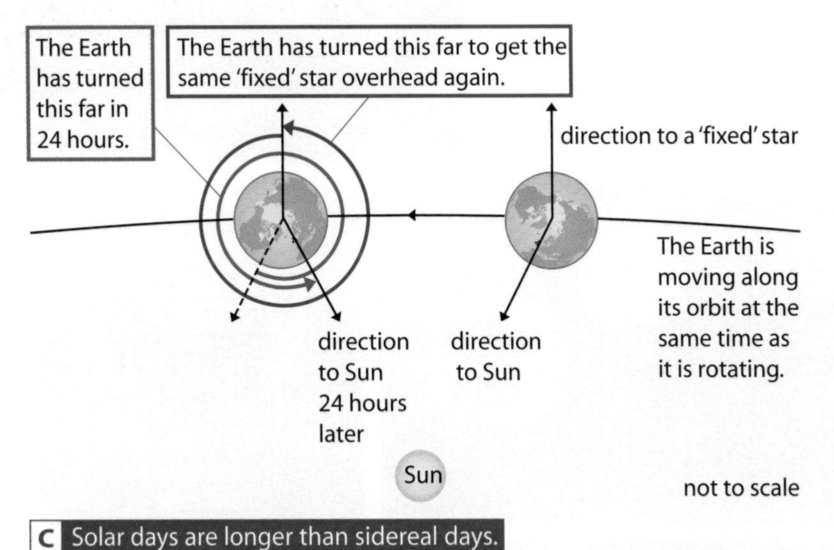

C Solar days are longer than sidereal days.

The Moon takes slightly longer than 24 hours to appear to move around the Earth once. This is because the Moon is moving in its orbit as the Earth is spinning.

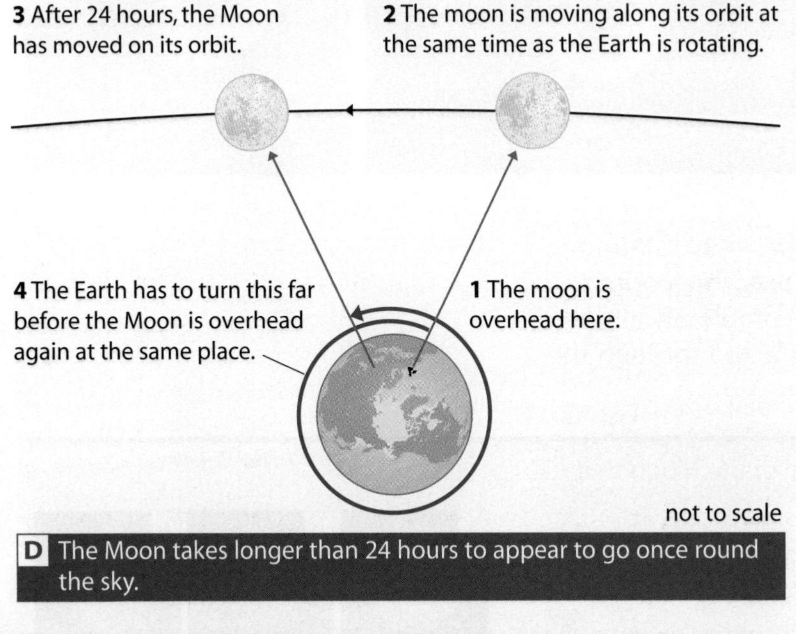

D The Moon takes longer than 24 hours to appear to go once round the sky.

H A sidereal day lasts 23 hours and 56 minutes, 4 minutes less than a solar day. It takes the Moon 24 hours and 49 minutes to appear to move across the sky once.

?

3 a What is a sidereal day?
 b What is a solar day?
 c Why is a solar day longer than a sidereal day?

4 Why does it appear to take the Moon longer than 24 hours to move across the sky once?

5 How many sidereal days are there in a year? Explain your answer.

Summary

Write short encyclopedia entries to explain the terms 'sidereal day' and 'solar day'. Include an explanation of why they are of different lengths.

Phases and eclipses

What causes the phases of the Moon and eclipses?

The Moon is a sphere, but it does not always look round from the Earth. The different shapes the Moon appears to have are called the **phases of the Moon**.

We see the Moon because light from the Sun is reflected from it. The Moon orbits the Earth, taking approximately 28 days for one orbit, so we cannot always see all of the side that is lit by the Sun. The side of the Moon away from the Sun is dark, and we cannot usually see it.

B The Moon at different stages during a lunar eclipse.

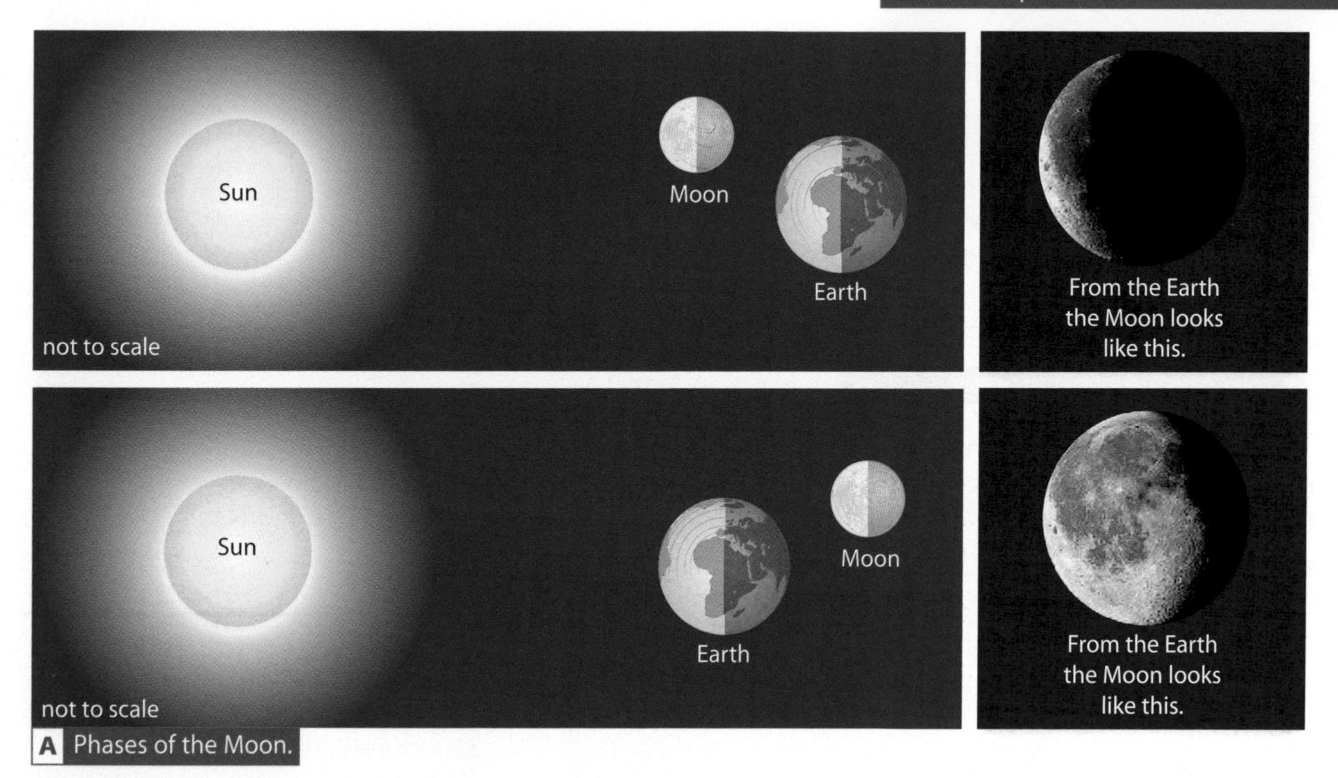

A Phases of the Moon.

As the Moon orbits the Earth, it sometimes moves into the shadow of the Earth. This is a **lunar eclipse**. The Moon is nearly always faintly visible even when completely in the shadow of the Earth, as some light is refracted through the Earth's atmosphere.

?

1. Why can't we always see all of the side of the Moon that faces us?
2. A full moon happens when the Moon looks like a complete circle. How long does it take from one full moon to the next?
3. Look at diagram A. Draw similar diagrams to show where the Moon is in its orbit when it looks like the photos in C.

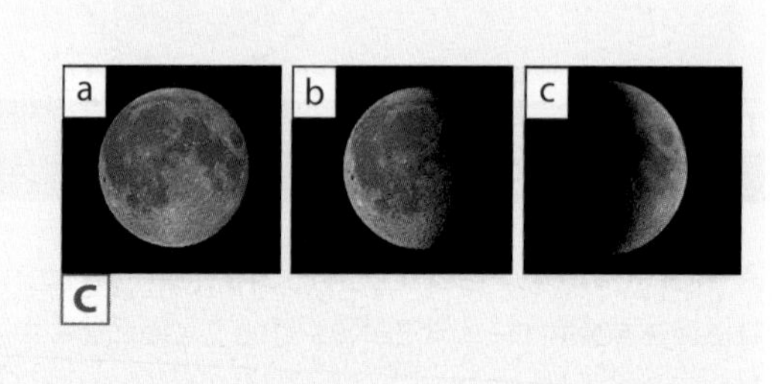

C

Sometimes the Moon comes between the Earth and the Sun. This is a **solar eclipse**. The Moon's shadow is much smaller than the Earth's surface, so solar eclipses can be seen only from the small area of the Earth that the Moon's shadow passes over. If observers see the Sun completely blocked by the Moon, they are seeing a **total eclipse**. If the Sun is only partly blocked out, they are seeing a **partial eclipse**.

D The shadow of the Moon on the Earth during a solar eclipse. People in the shadow see a total eclipse.

E A partial eclipse.

Sun

People here see a partial eclipse.

People here see a total eclipse.

not to scale

F How a solar eclipse occurs.

H You might expect that a solar eclipse would happen every month, because the Moon goes around the Earth once each lunar month. This does not happen, because the orbit of the Moon is tilted by about 5° relative to the orbit of the Earth.

?

4. Draw a diagram to show how a lunar eclipse happens.
5. Draw a diagram to show how a solar eclipse happens. Label on your diagram the places where people would see a total eclipse and where they would see a partial eclipse.
6. Explain why solar eclipses are visible from only a small part of the Earth, but lunar eclipses are visible from anywhere on the night side of the Earth.
7. Find out what an annular eclipse is, and why it happens.

Summary

Write glossary definitions for all the words in bold on these pages.

P7.3 Mapping the stars

How do astronomers record the positions of the stars?

If you are looking for a particular star, you need directions such as 'look due south, about 40° above the horizon'. This would tell you where to point your telescope. The direction is called the **azimuth**, and is usually given as a compass bearing (such as 20°), and the angle above the horizon is called the **altitude**. Unfortunately, this works only for one place at one particular time of day and time of year.

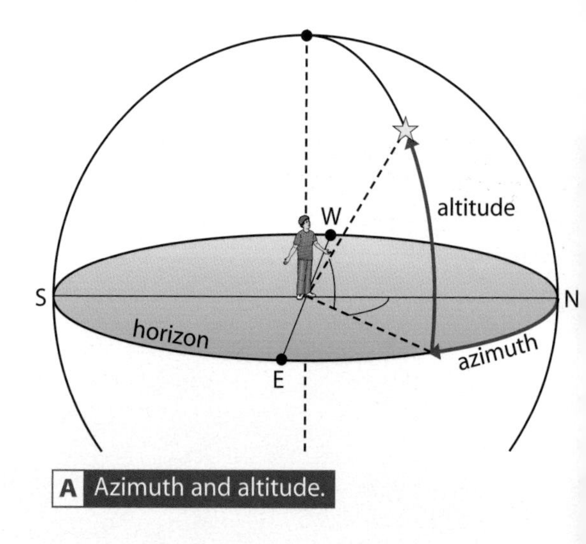

A Azimuth and altitude.

?

1 Why wouldn't azimuth and altitude directions work for:
  - a a different time of day
  - b the same time of day at a different time of year
  - c a different place?

Astronomers use the **equatorial system** of references, which is not affected by the movement of the Earth. Diagram B shows an imaginary transparent sphere around the Earth with grid lines on it. The **celestial equator** is a line above the Earth's equator, and the **celestial poles** are above the poles on the Earth. The distance above or below the celestial equator is called the **declination** or **celestial latitude**, and is measured in degrees. The distance around the sphere is called the **right ascension (RA)** or **celestial longitude**, and is measured from a fixed point called the **vernal equinox**. The right ascension can be measured in hours or in degrees, where 15° is equivalent to one hour.

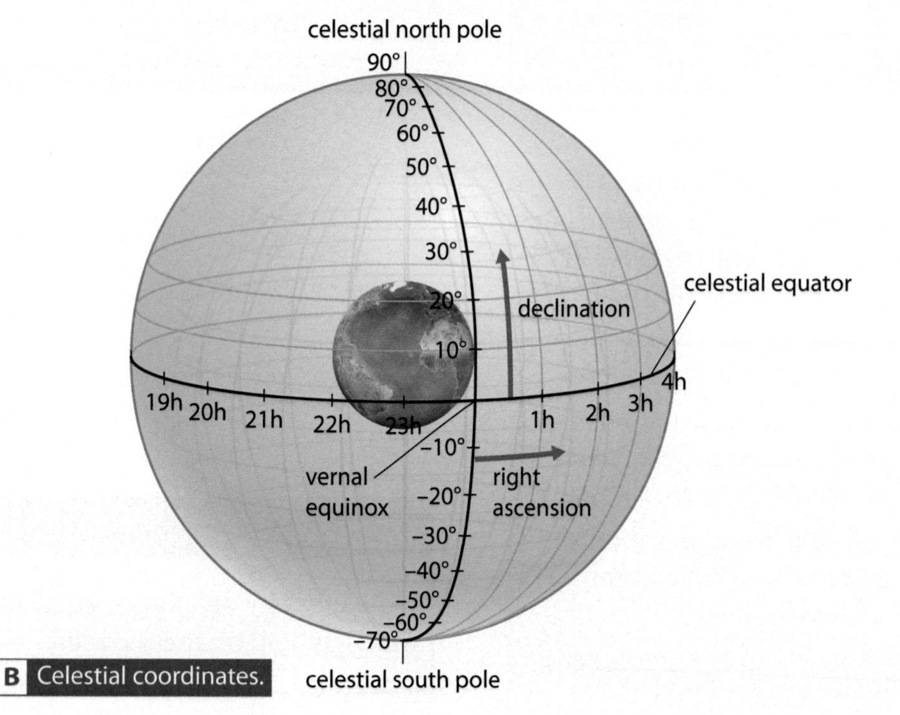

B Celestial coordinates.

Diagram C shows a star map with the declination and right ascension marked on it. The right ascension numbers go in the opposite direction to those in diagram B, because diagram C shows the grid as seen from the Earth.

?

2 Look at diagram C. Estimate the declination and right ascension of:
   a Bellatrix
   b Rigel.

3 Where will the constellation Orion be at 9 pm? Explain your answer.

The stars near the north celestial pole are visible from the UK throughout the year. Photo A on page 114 shows the stars appearing to move around the north celestial pole. The only bright star that appears to move hardly at all is Polaris, sometimes called the Pole Star, which is very close to the north celestial pole. Other stars are visible only at certain times of the year.

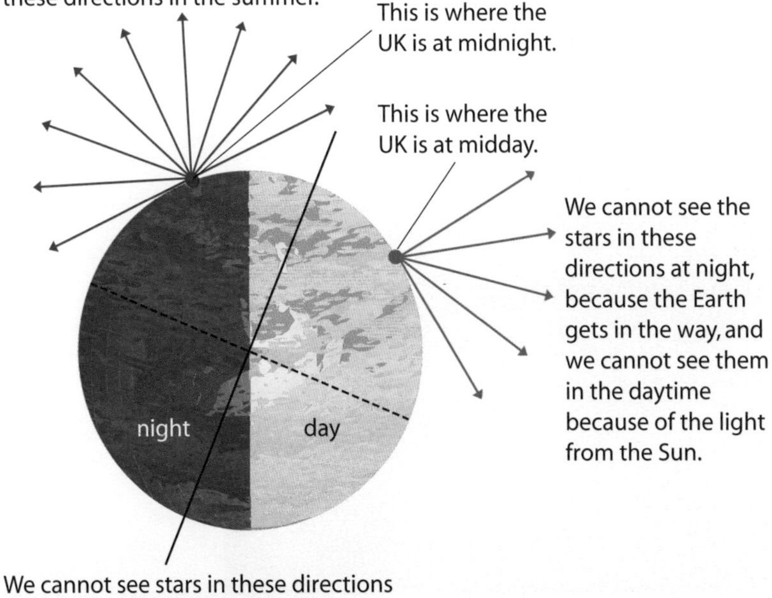

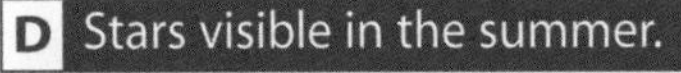
D Stars visible in the summer.

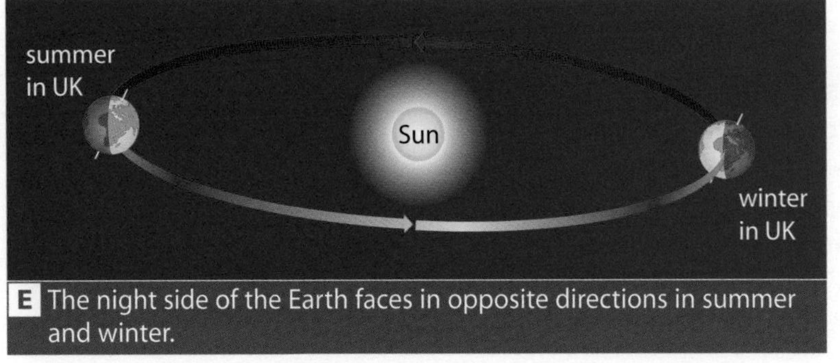

E The night side of the Earth faces in opposite directions in summer and winter.

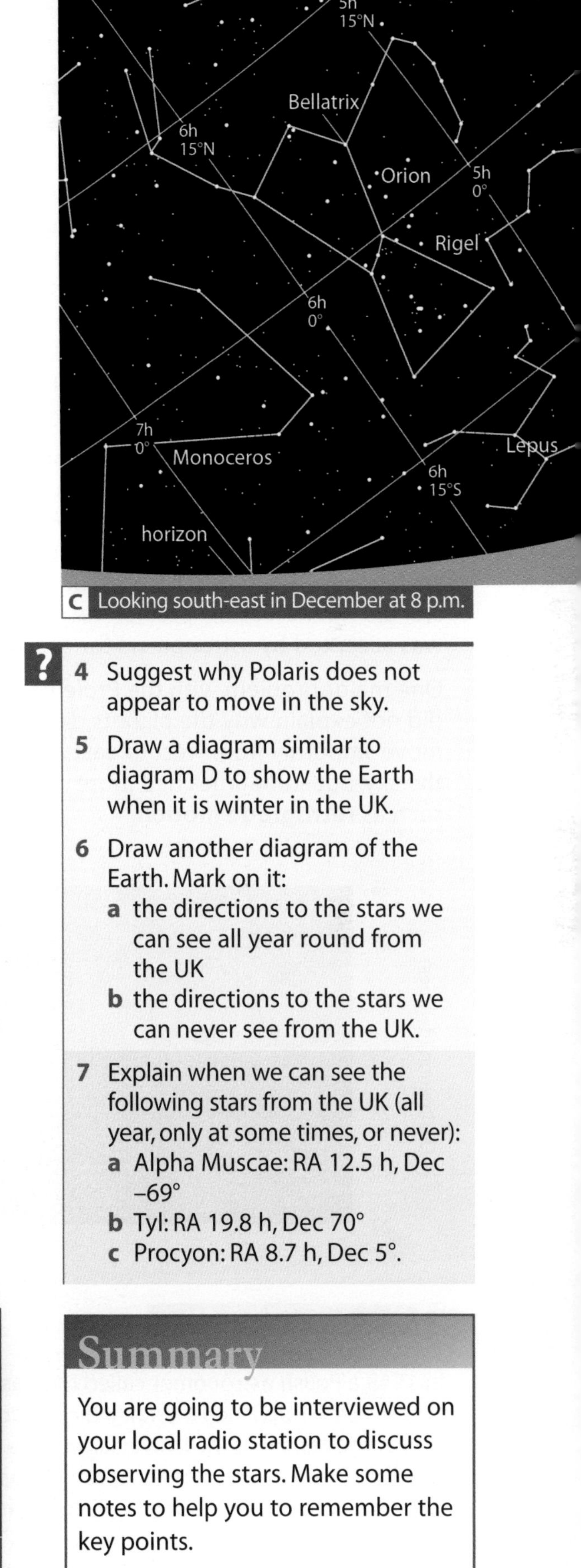

C Looking south-east in December at 8 p.m.

?

4 Suggest why Polaris does not appear to move in the sky.

5 Draw a diagram similar to diagram D to show the Earth when it is winter in the UK.

6 Draw another diagram of the Earth. Mark on it:
   a the directions to the stars we can see all year round from the UK
   b the directions to the stars we can never see from the UK.

7 Explain when we can see the following stars from the UK (all year, only at some times, or never):
   a Alpha Muscae: RA 12.5 h, Dec −69°
   b Tyl: RA 19.8 h, Dec 70°
   c Procyon: RA 8.7 h, Dec 5°.

Summary

You are going to be interviewed on your local radio station to discuss observing the stars. Make some notes to help you to remember the key points.

P7.4 Movements of the planets

How do the planets seem to move against the stars?

There are five planets in our Solar System that can be observed with the naked eye: Mercury, Venus, Mars, Jupiter and Saturn. For thousands of years astronomers made careful observations of the positions of these planets against the stars, and used their observations to try to work out how the Sun and planets were arranged. Most early ideas had the Earth at the centre of the Universe. One such idea was published by an Egyptian astronomer called Ptolemy in the 2nd century CE.

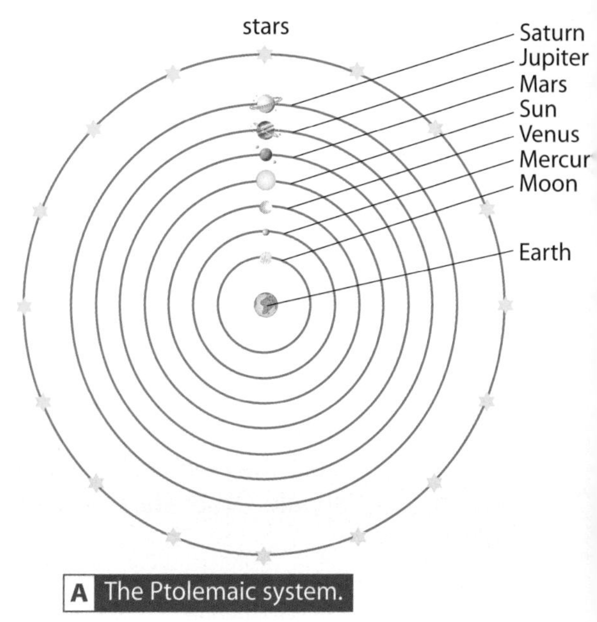

A The Ptolemaic system.

The Ptolemaic model had the Sun and all the planets moving in circular orbits around the Earth. Unfortunately, the paths of the planets predicted by this model did not exactly match the way the planets appeared to move, so Ptolemy added smaller circles and other modifications. Although the model still did not accurately predict the motions of the planets, it was accepted by astronomers for over 1500 years.

One major problem with the Ptolemaic model was that it did not explain why the planets did not always appear to move smoothly from west to east across the fixed stars in the sky, but sometimes had more complicated movements such as **retrograde motion**.

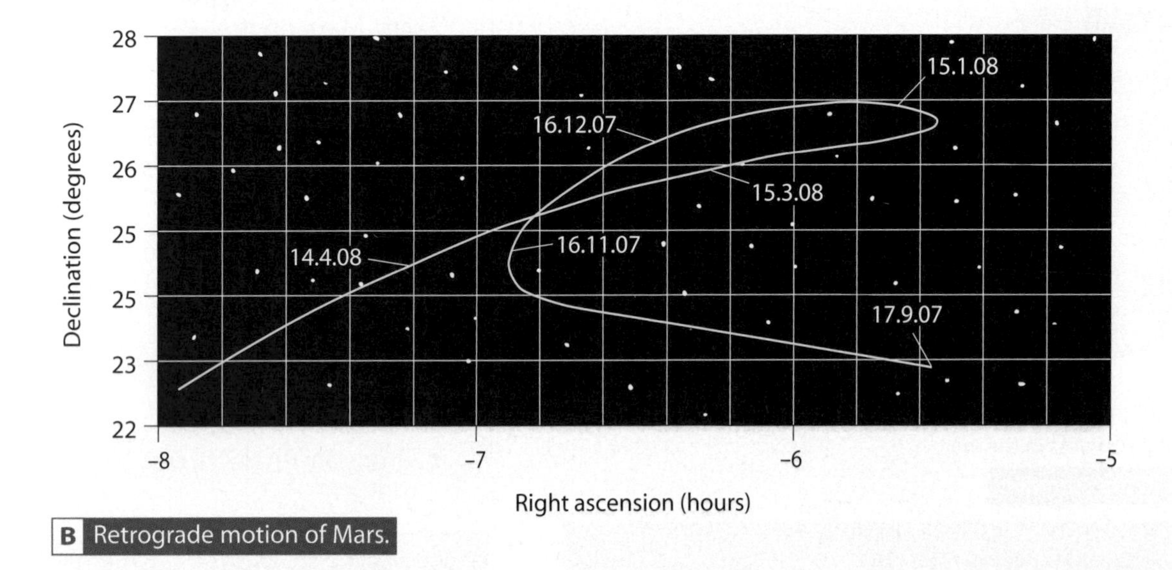

B Retrograde motion of Mars.

In 1543 a Polish astronomer called Nicolas Copernicus (1473–1543) published a book which suggested that the Sun was at the centre of the Universe, and the Earth moved around the Sun. This idea was not accepted at the time, but further work by astronomers such as Galileo Galilei (1564–1642), Tycho Brahe (1546–1601) and Johannes Kepler (1571–1630) led to the model of the Solar System that astronomers accept today.

?

1 Look at diagram B.
- **a** Which way do planets normally appear to move across the fixed stars?
- **b** What is retrograde motion?

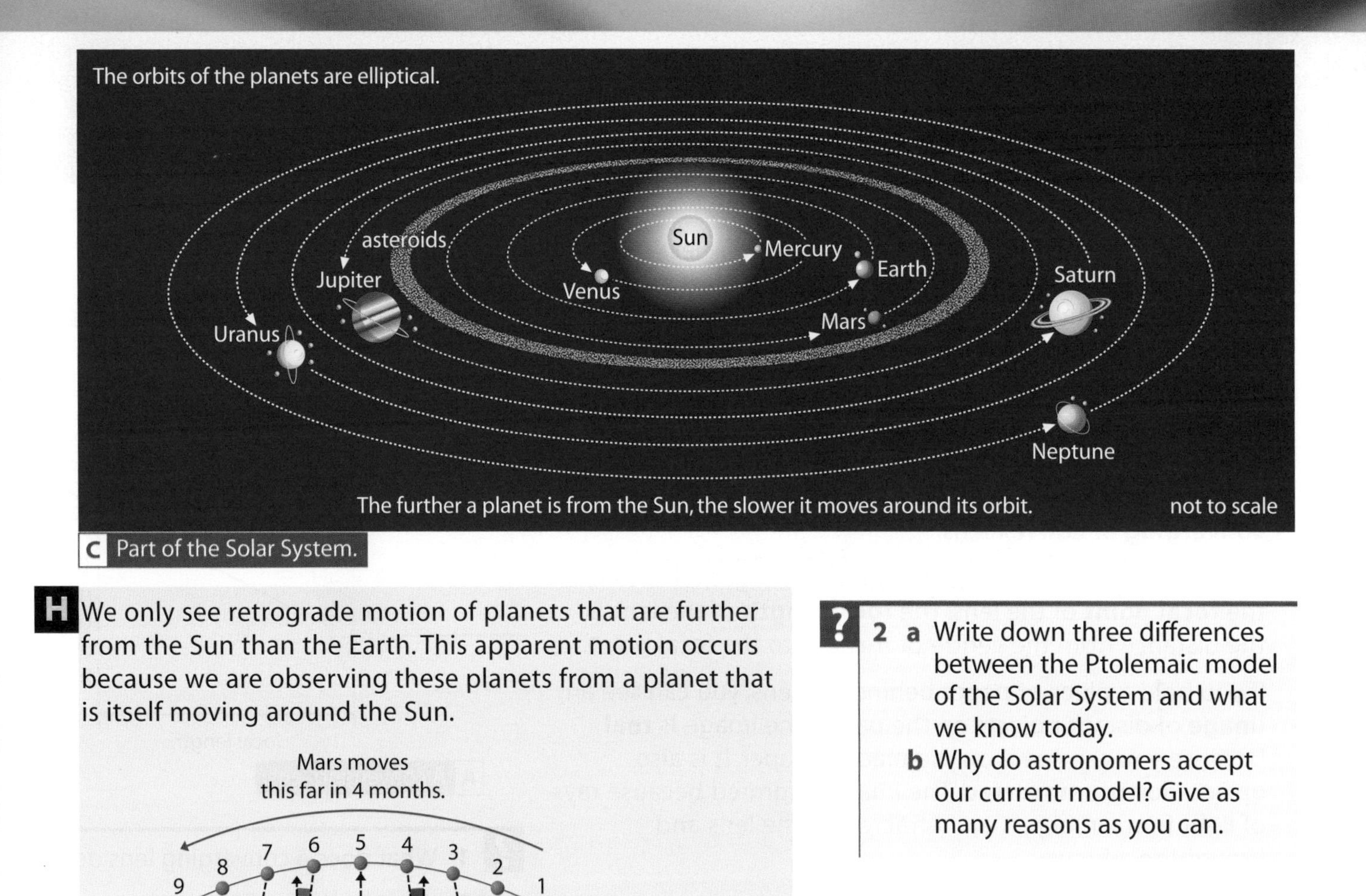

C Part of the Solar System.

H We only see retrograde motion of planets that are further from the Sun than the Earth. This apparent motion occurs because we are observing these planets from a planet that is itself moving around the Sun.

D Observing Mars from the Earth.

?

3 Look at diagram D. Mars appears to be moving from west to east against the fixed stars between times 1 and 2. How does it appear to be moving between all the other pairs of times on the diagram?

4 Sketch a diagram similar to diagram D to show how the position of Venus would appear to move. (*Hint*: remember that Venus is closer to the Sun than the Earth.)

5 Venus is sometimes called the 'morning star' or the 'evening star'. Find out why it was called this, and how this can be explained using our model of the Solar System.

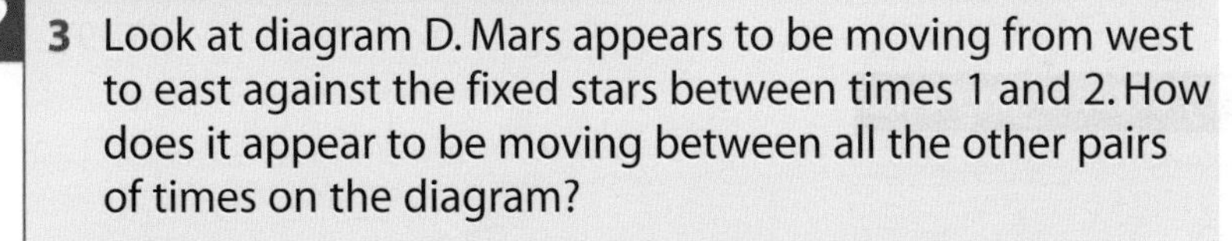

?

2 a Write down three differences between the Ptolemaic model of the Solar System and what we know today.

b Why do astronomers accept our current model? Give as many reasons as you can.

Summary

Write a short paragraph to answer the question at the top of page 120.

H Explain *why* the planets seem to move as they do.

P7.5 Lenses

What do lenses do?

Lenses are used in many things, such as spectacles, microscopes, telescopes, cameras and film projectors. Lenses are made of transparent materials like glass. Light is **refracted** when it goes into a lens and again when it comes out of the lens.

The lens in diagram A is fatter in the middle than at the edges. The lens makes the parallel light rays **converge** (come together) at one point. This type of lens is called a **converging** or **convex** lens.

The point at which the rays of light come together is called the **focal point** of the lens. The **focal length** of the lens is the distance from the centre of the lens to the focal point.

If you hold a piece of paper behind the lens, you can see an **image** of distant objects on the paper. The image is **real** because you can see it on the piece of paper. It is also upside-down, or **inverted**. The image is formed because rays of light from an object are refracted by the lens and brought to a focus.

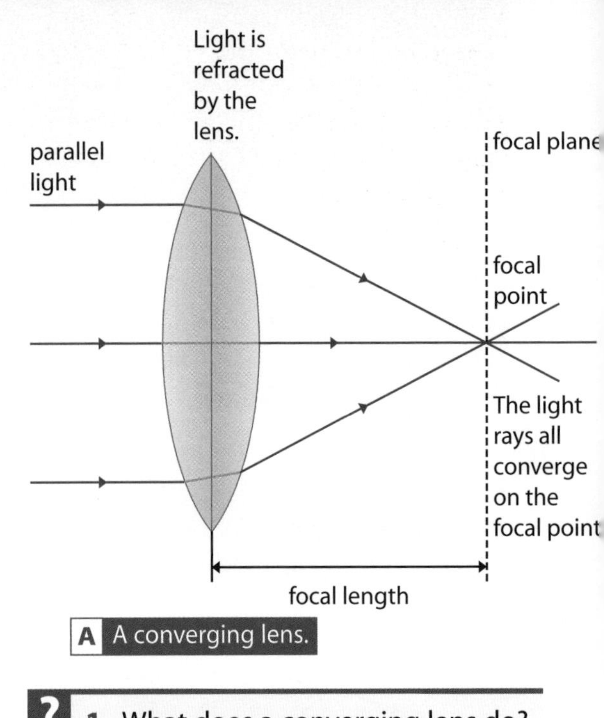

A A converging lens.

?

1 What does a converging lens do?
2 What is the focal point of a lens?
3 What is the focal length of a lens?

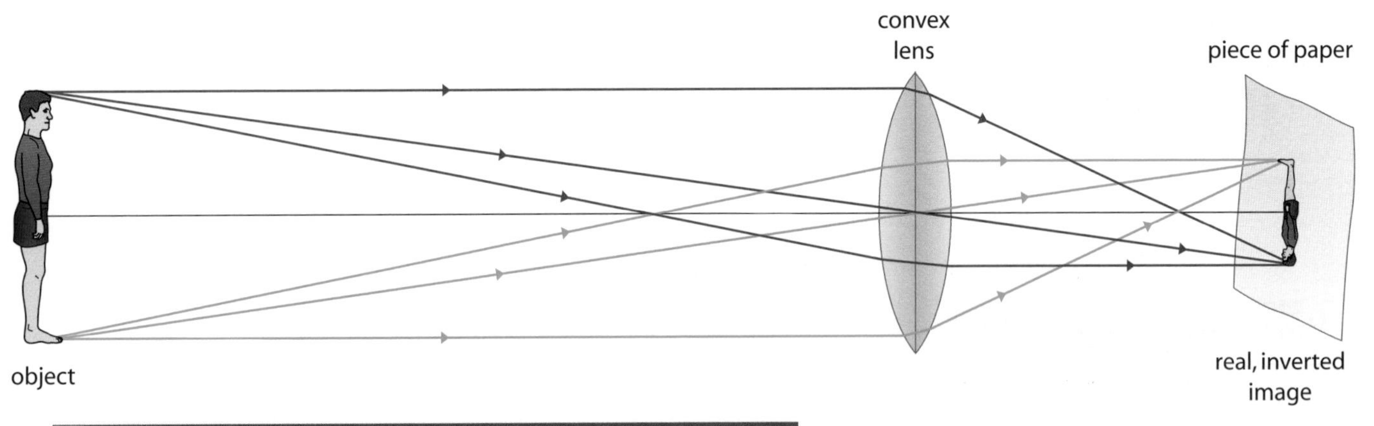

C How an image of a distant object is formed by a converging lens.

We see most objects because they reflect light. The reflected light is scattered in all directions. If you are close to the light source, you can see that there is an angle between the rays, but as you move further and further away from the source, the angle between the rays becomes smaller and smaller. When you are a long way from a light source, the angle between the rays is so small that the rays are effectively parallel. The image formed by a converging lens from parallel rays of light is formed on the focal plane of the lens.

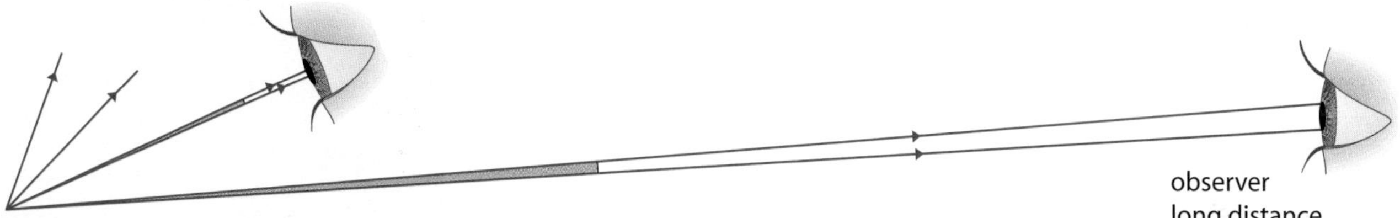

observer close to source

point source

observer long distance from source

D Rays of light from a very distant source are effectively parallel.

?

4 You use a converging lens to form an image of the Sun on a piece of paper. Where is the image formed?

Thicker lenses have more curved surfaces than thinner lenses, so they refract light more. This means that the focal point is closer to the lens and so the focal length is shorter.

E We can work out the power of a lens using this equation:

$$\text{power (in dioptres)} = \frac{1}{\text{focal length of lens (in metres)}}$$

The focal length of a lens is 200 cm. What is its power?
Convert the units of length to metres:
200 cm = 2 m
power = $^1/_2$ = 0.5 dioptre

?

5 Why is a thicker lens more powerful than a thin lens?

6 Work out the power of the following lenses:
a focal length = 1.25 m
b focal length = 400 mm.

7 Lens A has a focal length of 0.5 m. Lens B has a power of 2.5 dioptres. Which lens is more powerful?

Summary

Draw and label a diagram to show how a converging lens forms an image of a distant object.

P7.6

Simple telescopes

How does a simple telescope work?

The rays of light emitted by a luminous source, such as a star, travel away from the source in all directions.

Stars are very big objects, but they are so far away that the rays of light from them are effectively parallel when they reach us. They look like just points of light.

?

1 What is a luminous source?

2 Why is the light from stars effectively parallel?

To study stars in more detail, the image has to be made larger, or magnified. A **telescope** can be used to look at stars in more detail.

A simple telescope is made of two converging lenses. Light enters the telescope through the **objective** lens and forms an image on the focal plane of the lens. You look through a second lens, which is called the **eyepiece**. The eyepiece lens is more powerful than the objective lens. The distance between the two lenses is the sum of the focal lengths of the two lenses. When you look through the eyepiece, you see an inverted image of the object.

A Telescopes can also be used to look at the planets in our Solar System. This photo of Mars was taken using the Hubble Space Telescope.

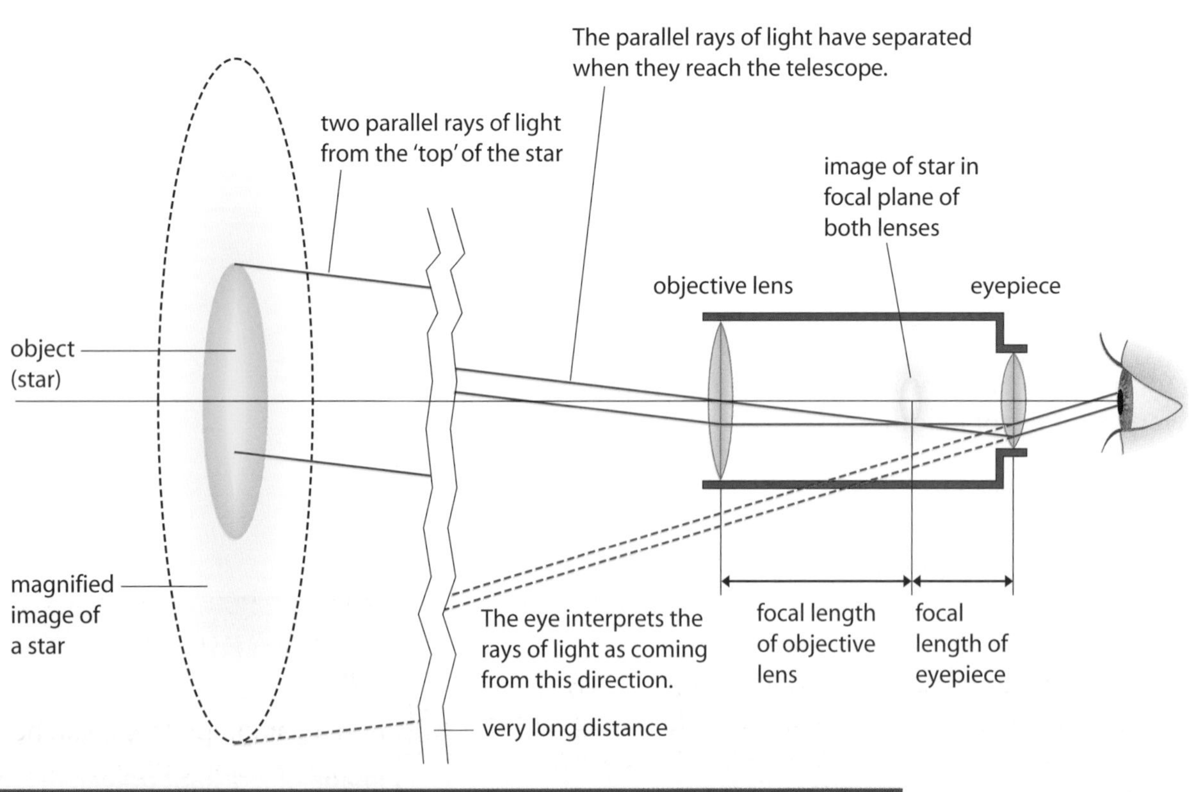

B In a simple telescope, rays from the top of the object form the bottom of the image.

3 Through which lens does light enter a telescope?

4 A telescope is made using two lenses with focal lengths 60 cm and 10 cm. How far apart should the two lenses be?

5 Describe the image formed by a telescope.

H The angle formed by the image at the eye is bigger than the angle formed by the object.

E The angular magnification can be calculated using this equation:

$$\text{angular magnification} = \frac{\text{focal length of objective lens (in metres)}}{\text{focal length of eyepiece lens (in metres)}}$$

The magnification is a ratio of two lengths, so it does not have units.

The focal lengths of the objective and eyepiece lenses in a telescope are 0.8 m and 0.2 m respectively.

$$\text{angular magnification} = \frac{0.8}{0.2} = 4$$

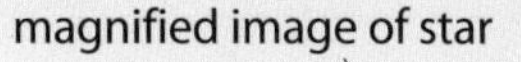
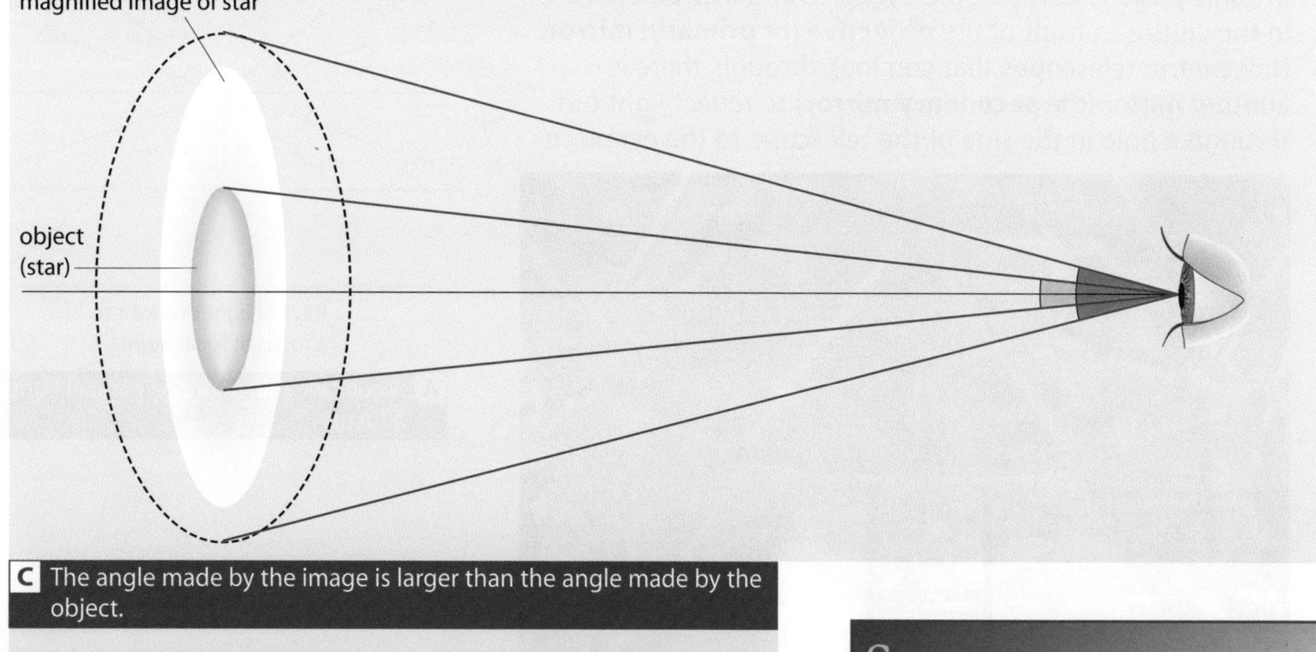

C The angle made by the image is larger than the angle made by the object.

6 In a telescope, the objective lens has a focal length of 0.5 m and the eyepiece has a focal length of 0.1 m. What is the angular magnification of the telescope?

7 Which combination of lenses would produce the biggest angular magnification? Choose from the following lenses: lens A has a focal length of 1 m; lens B has a focal length of 25 cm; lens C has a power of 5 dioptres.

Summary

Write two brief encyclopaedia entries to describe how a telescope works.

H Add a paragraph to explain how its angular magnification is calculated.

P7.7

Astronomical telescopes

Why are mirrors used in telescopes?

As stars are very distant objects, not only is the light from them parallel, but it is also very faint. During the day, we can see light from the Sun only because it is so much stronger than light from the stars. To take a good photo of a star, we need to collect as much light as possible, without light from other sources spoiling the image.

In low light levels, the **aperture** of a telescope needs to be as wide as possible to allow more light in. This is similar to the way the pupil of your eye expands in dim light.

In many telescopes, the objective is several metres across. It is very difficult to make lenses this big. The glass would have to be very pure and the lenses would be incredibly expensive.

Most astronomical telescopes use a **concave** mirror to collect light. A concave mirror has the same effect as a converging lens. It brings parallel rays of light to a focus to form a real, inverted image.

In some large telescopes, the eyepiece and instruments are in the centre, in front of the **objective** (or **primary**) **mirror**. However, in telescopes that you look through, there is another mirror (the **secondary mirror**) to reflect light out through a hole in the side of the telescope to the eyepiece.

B This large primary mirror is made up of lots of smaller hexagonal mirrors.

?

1. Why does a telescope need a large aperture?
2. Why are lenses not used for the objective in most astronomical telescopes?

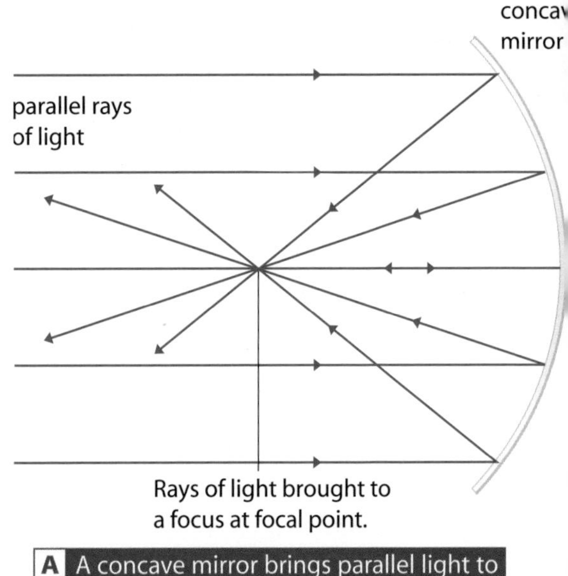

A A concave mirror brings parallel light to a focus.

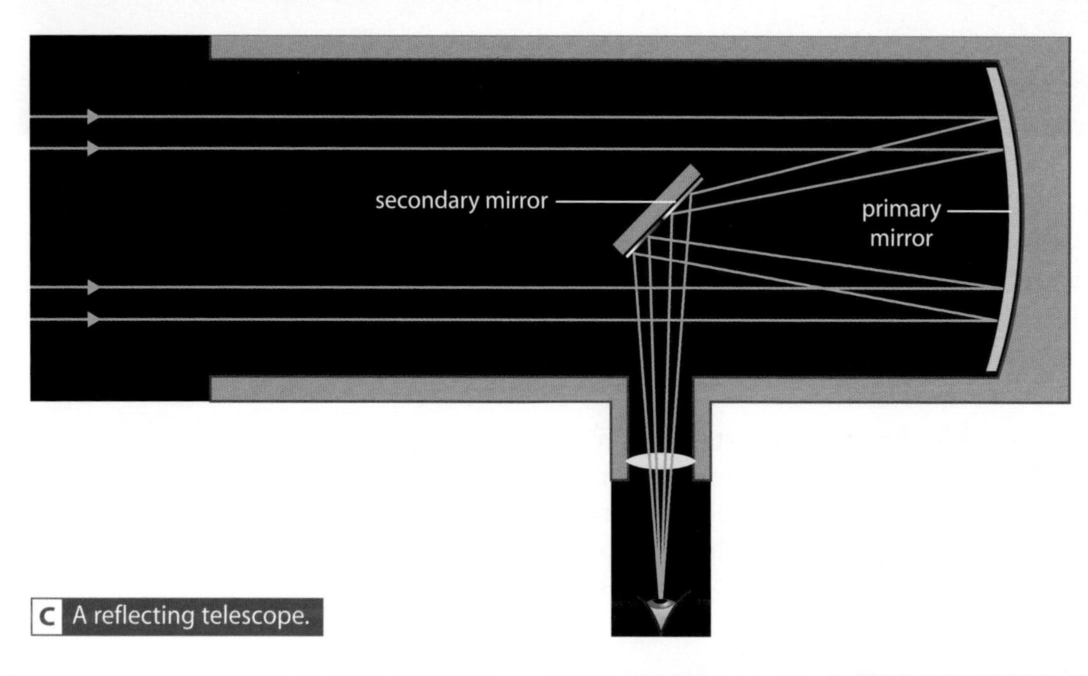

C A reflecting telescope.

H When light goes through the aperture, the edges of the wave bend slightly. This is known as **diffraction**. If the size of the aperture is similar to the wavelength of the light, there is a lot of diffraction. This means that the light hitting the concave mirror is no longer parallel, so the mirror does not produce a sharp image.

If the size of the aperture is much bigger than the wavelength of the light, the effects of diffraction are small and the image is sharp.

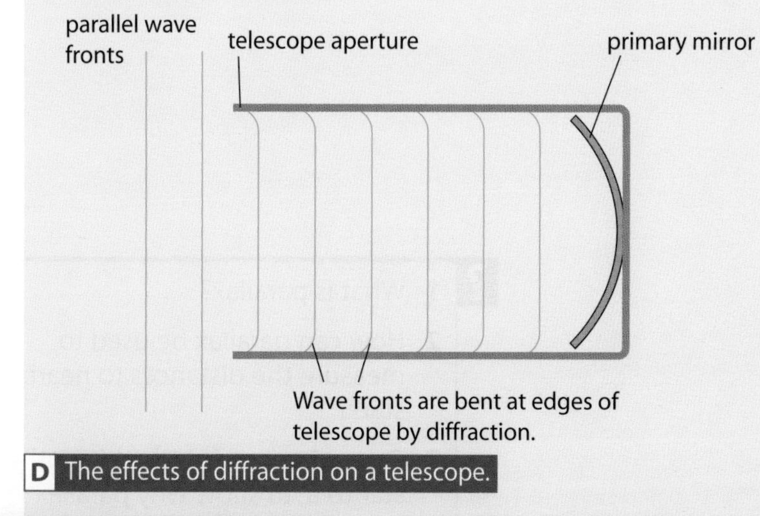

D The effects of diffraction on a telescope.

?

6 What is diffraction?

7 Why does the aperture of a telescope need to be much bigger than the wavelength of the light you are observing?

8 Find out what other advantages a mirror has over a lens. (*Hint*: find out what happens when light of different wavelengths is refracted.)

?

3 What does a concave mirror do in a telescope?

4 List the similarities and differences between a converging lens and a concave mirror.

5 Why do you think the light has to be reflected by a secondary mirror?

Summary

Copy and label the diagram of a reflecting telescope and add notes to explain its key features.

P7.8 Parallax

How can astronomers measure distances to the stars?

Parallax is the apparent movement of objects at different distances from an observer. You can see this effect for yourself if you look out of the window of a moving bus or train; although objects in the landscape are stationary, things close to you appear to be moving backwards faster than more distant objects.

A similar effect can be seen in the stars, with stars closer to the Earth appearing to move against a background of more distant stars as the Earth moves around its orbit. Parallax in the stars is much too small to be detected with the naked eye; it was not observed until suitable telescopes were built.

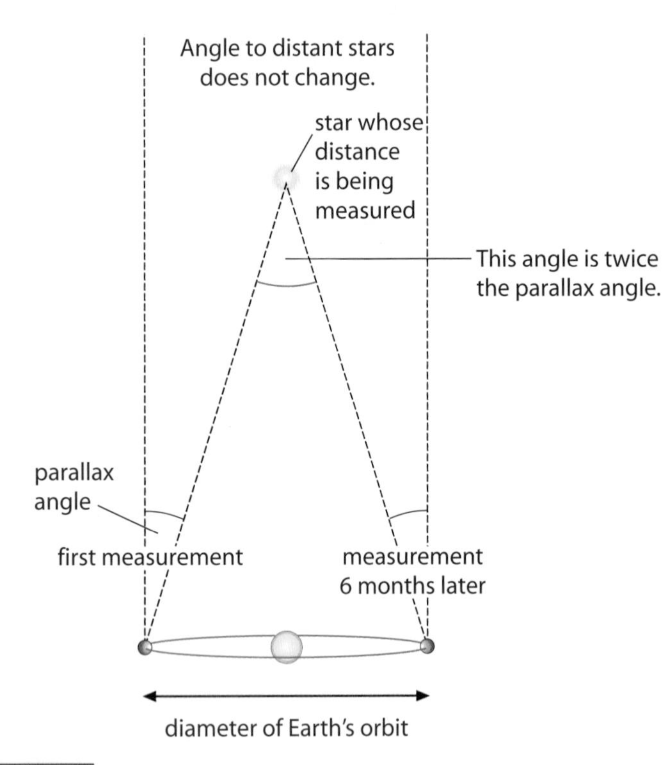

A Parallax.

The **parallax angles** of even the closest stars are very small, and are measured in small angular units called **arcseconds**. A degree is divided into 60 minutes of arc, and each minute of arc is divided into seconds of arc, or arcseconds. One arcsecond is therefore 1/3600 of a degree, and is sometimes shown as ". The parallax angle of Proxima Centauri, the nearest star to the Earth, is only 0.772 arcseconds (or 0.772"). More distant stars have even smaller parallax angles.

?

1. What is parallax?
2. How can parallax be used to measure the distances to nearby stars?
3. Copy diagram A and add another star to it, to show why parallax angles are smaller for more distant stars.
4. Why do you think astronomers make parallax measurements six months apart, instead of just measuring at the beginning and end of the night?

Using the equation given in diagram B, the distance of a star with a parallax angle of 1 arcsecond is:

$$\frac{149\,600\,000\,000 \text{ m}}{\tan(0.000\,277\,78°)}$$

$$= 3.086 \times 10^{16} \text{ m}$$

$0.000\,277\,78° = 1/3600°$

Because this is such a large number, astronomers use other units to measure the distances to stars. One such distance is the **light year**. This is the distance that light travels in a year, and is approximately 9.467×10^{15} m. An alternative unit is the **parsec** (pc), which is the distance of a star whose parallax angle is 1 arcsecond (the name is an abbreviation of '*par*allax *sec*ond'). A parsec therefore equals 3.086×10^{16} m. Using this unit allows us to simplify the formula for calculating the distance to a star.

E The distance to a star (in parsecs) can be calculated from the parallax angle (in arcseconds):

$$\text{distance (parsecs)} = \frac{1}{\text{parallax (arcseconds)}}$$

The nearest star, Proxima Centauri, has a parallax angle of 0.772 arcseconds. How far is it from the Earth in parsecs?

$$\text{distance (parsecs)} = \frac{1}{\text{parallax (arcseconds)}}$$

$$= \frac{1}{0.772''}$$

$$= 1.295 \text{ parsecs}$$

?

5 Why are parsecs or light years normally used to measure distances to stars?

6 The table shows the parallax angles of three stars.

Star	Parallax angle
Rigel	0.004"
Bellatrix	0.013"
Polaris	0.007"

a How can you tell from the parallax angles which of these stars is closest to Earth?
b Calculate the distances to these stars in parsecs.

7 Calculate the distances to the stars in question **6** in metres.

8 Sirius is 8.7 light years away.
a How far is this in parsecs?
b What parallax angle does it have?

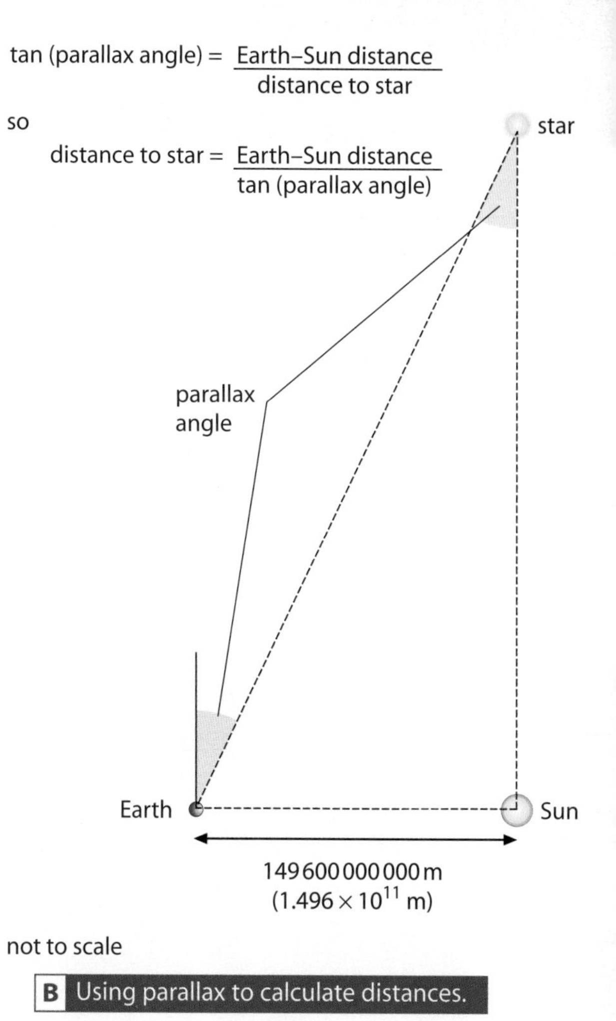

B Using parallax to calculate distances.

Summary

Draw and label a diagram to explain how parallax is used to measure the distances to nearby stars.

P7.9 Light from the stars

What can the light emitted from stars tell us?

An observer without a telescope can see that some stars are brighter than others, and in some cases can also see that stars are not all the same colour. However, with the invention of accurate telescopes, and photographic plates to record observations, light from the stars could be analysed in far more detail than is possible with the naked eye.

All hot objects emit electromagnetic radiation over a continuous range of frequencies. The brightness of a star depends on its temperature and its size – bigger stars have a bigger surface area to emit electromagnetic radiation. Bigger stars are also hotter.

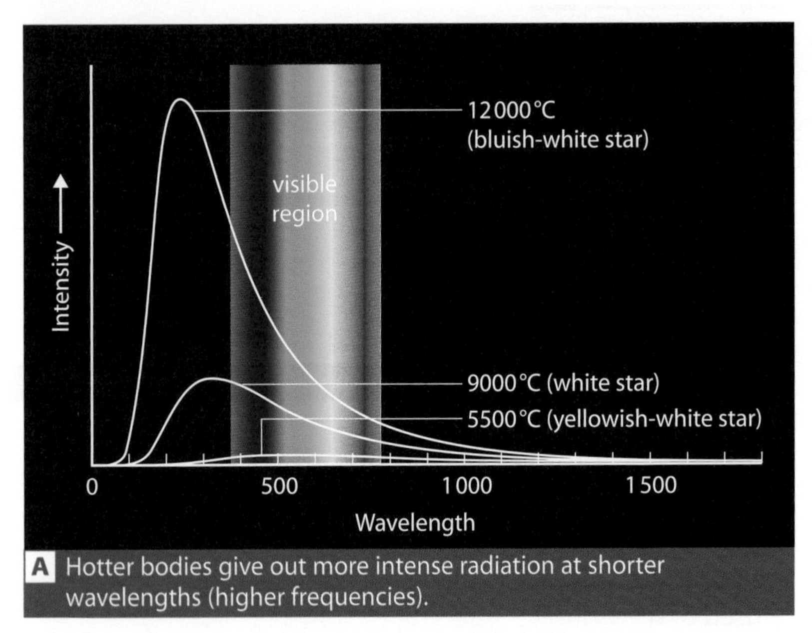

A Hotter bodies give out more intense radiation at shorter wavelengths (higher frequencies).

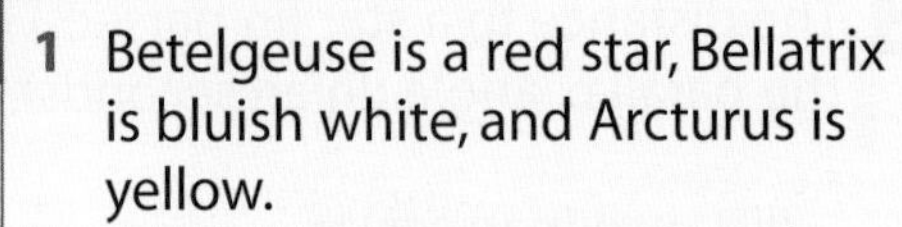

?

1 Betelgeuse is a red star, Bellatrix is bluish white, and Arcturus is yellow.
   - **a** Put these stars in order of temperature, with the hottest first.
   - **b** Explain how you worked out your answer.

2 What further information do you need to work out which star in question **1** is the brightest? Explain your answer.

Light from the Sun or the stars can be examined using a prism to split it up into separate colours. When this is done, black lines appear across the spectrum corresponding to particular wavelengths of light absorbed by the gases surrounding the star. A spectrum showing these black lines is called an **absorption spectrum**. Each pattern of lines corresponds to different elements in the gas absorbing energy, and the patterns of lines can be used to identify elements in the stars.

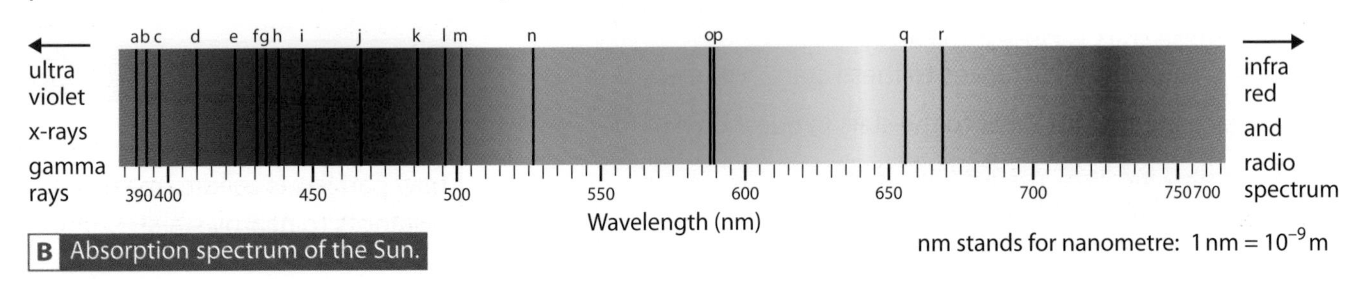

B Absorption spectrum of the Sun.

nm stands for nanometre: $1\ \text{nm} = 10^{-9}\ \text{m}$

What causes the lines?

The surface temperatures of stars vary from around 2000 °C to 60 000 °C. These temperatures are hot enough to **ionise** some atoms by allowing electrons to escape from them. In some cases, electrons do not escape from the atoms but change to a higher energy state. Diagram C is an **energy level diagram** illustrating some of the energy states for hydrogen. A hydrogen atom can absorb energy only in certain amounts, represented by the arrows on the diagram. Each amount of energy absorbed corresponds to a particular wavelength of electromagnetic radiation missing from the spectrum, and shows up as a black line.

Each element has its own particular electron structure, and so can absorb different wavelengths of radiation. Radiation from a star passes through cooler gases in the atmosphere of the star, and the atmosphere absorbs certain wavelengths of light according to the elements present. Absorption lines are seen at other wavelengths of the electromagnetic spectrum in addition to the wavelengths of visible light.

The absorption spectrum from a star contains lines from many different elements, but comparison with spectra obtained from single elements in a laboratory on Earth allows the different elements to be identified. The table shows some of the wavelengths at which different elements absorb light.

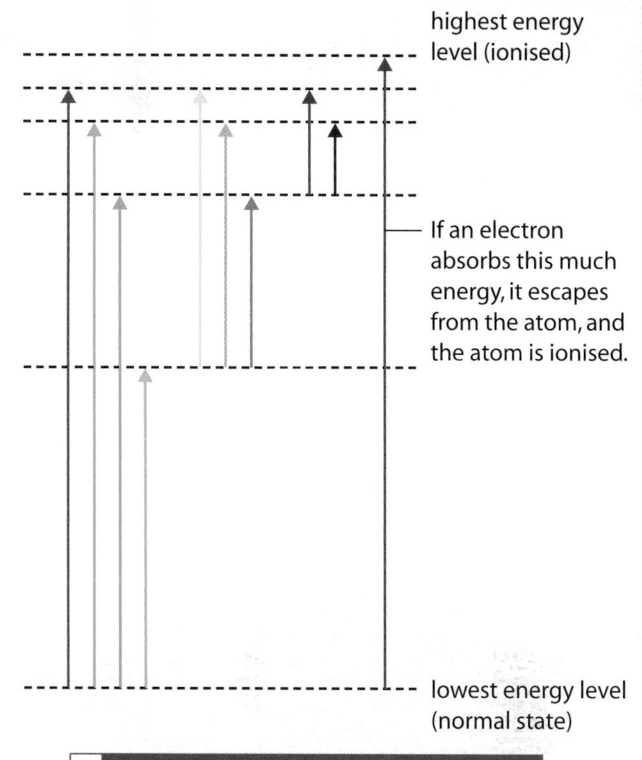

C An energy level diagram for hydrogen.

Element	Wavelengths (nm)
Ca	393, 397, 423, 431
H	410, 434, 486, 656
Fe	431, 438, 467, 496, 527
Na	589, 590
He	389, 447, 502, 588, 668

D Wavelengths of light absorbed by some elements.

?

3 What is an absorption spectrum?

4 What can absorption spectra tell us about the stars?

5 What causes the dark lines in an absorption spectrum?

6 Which lines in diagram B are due to the following elements?
- **a** hydrogen
- **b** sodium

7 Light from the Moon and from Mars has the same spectral lines as light from the Sun. Explain this statement.

Summary

Make a list of all the things the light from a star can tell us, and give a brief explanation for each point.

P7.10 The brightness of stars

How do astronomers use brightness to measure distance?

The parallax method described in Topic P7.8 depends on very accurate measurements of very small angles. The most accurate measurements of parallax to date were made by the European Space Agency's Hipparcos satellite between 1989 and 1993. The mission measured the angles to more than 100 000 stars far more accurately than ever before, and allowed the distances to stars up to 650 parsecs away to be measured accurately.

The Milky Way is at least 30 kiloparsecs across, so parallax cannot be used to measure distances to most of the stars in our own galaxy, let alone the distances to other galaxies. One alternative method of measuring distance relies on measuring the brightness of stars.

?

1 **a** What is the furthest distance at which parallax measurements provide reasonable accuracy?
b Why is the parallax method limited to relatively close stars?

If you look at a light source, such as a light bulb, its **apparent brightness** (how bright it looks to you) depends on:

- its **intrinsic brightness** (how bright it really is when viewed from close by)
- how far away it is.

This means that if astronomers know the intrinsic brightness of a star, they can use its apparent brightness to work out how far away it is. The intrinsic brightness of a star depends on its temperature and its size. The temperature can be estimated from the colour of the star. However, measuring the size of a star is even harder than measuring the distance to it.

?

2 Why does the distance to a light source affect its apparent brightness?

3 Why is measuring the size of a star harder than measuring the distance to it?

There are many types of stars, including some **variable stars** that change in brightness. **Cepheid variables** are stars that change in brightness with a regular **period**.

At this distance from the star, the light energy it emits is spread out over a sphere this size.

As the light travels further, it spreads out over an even greater area. A sphere of twice the radius has four times the surface area, so as the distance doubles the apparent brightness goes down by a factor of 4.

A

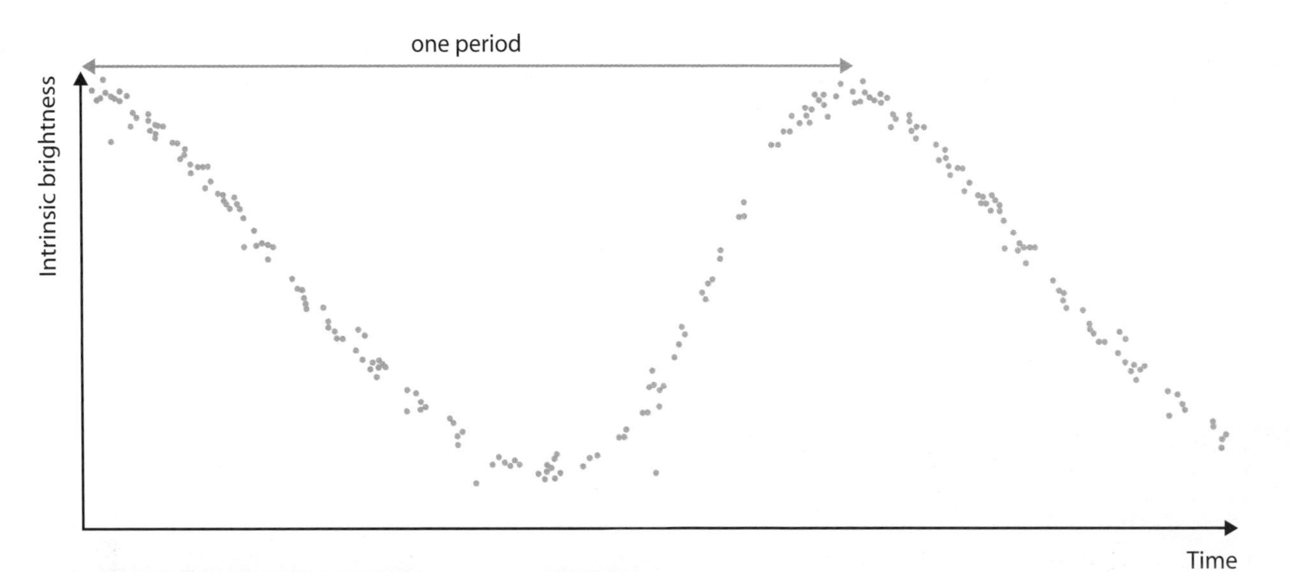

B Intrinsic brightness and period for a Cepheid variable.

In 1912, Henrietta Leavitt was analysing photographs of stars in the Small Magellanic Cloud, which is a small galaxy near the Milky Way. She discovered a relationship between the period of a Cepheid variable and its apparent brightness – the longer the period, the brighter the star. As all the stars she was studying were in the same galaxy, it could be assumed that they were all at approximately the same distance from the Earth. This meant that any differences in apparent brightness were due to differences in intrinsic brightness.

The Small Magellanic Cloud is too far away for accurate distance measurements to be made using parallax, but a number of other methods were used to estimate its distance. Once astronomers knew this, they could work out the intrinsic brightness of Cepheid variables with different periods.

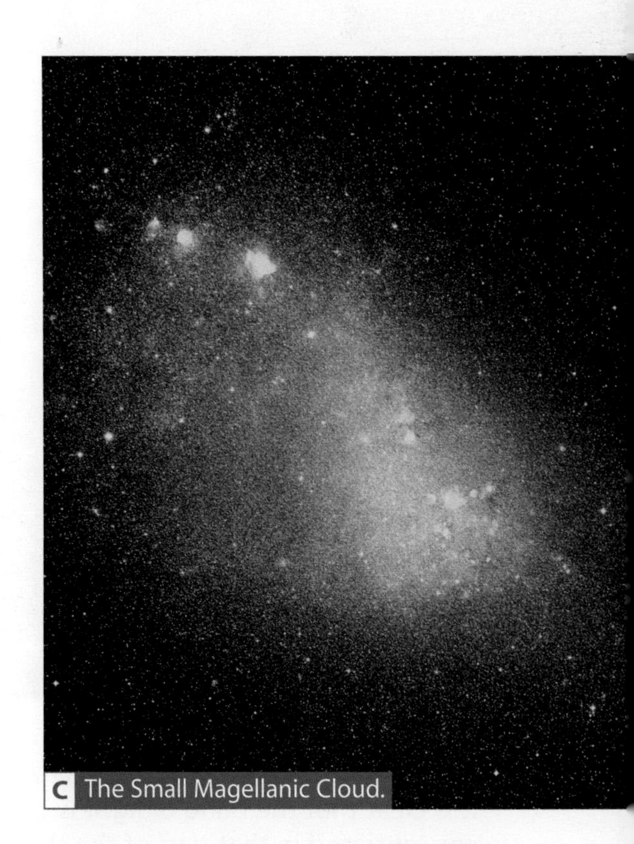
C The Small Magellanic Cloud.

?

4 **a** What does the 'period' of a star mean?
b Sketch a copy of graph B, and add a line to it showing the variations in brightness of a Cepheid variable with a shorter period.

5 Why is graph B shown as a series of dots rather than a line?

6 A Cepheid variable is discovered, and you want to know its distance.
a What two things would you need to measure?
b How can the distance be worked out from these measurements?

7 Cepheid variables are used as 'standard candles'. What do you think this means?

Summary

Make up a 'fill in the gaps' exercise to summarise the work on these pages.

P7.11 Nebulae and galaxies

How were other galaxies discovered?

When we look at the night sky we see stars as points of light. There are also a few **nebulae**, or patches of fuzzy light that are visible to the naked eye and have been known about since ancient times. When viewed with a telescope, some of these fuzzy patches could be resolved into clusters of stars, but some still looked like clouds of gas. As telescopes improved, more and more nebulae were discovered.

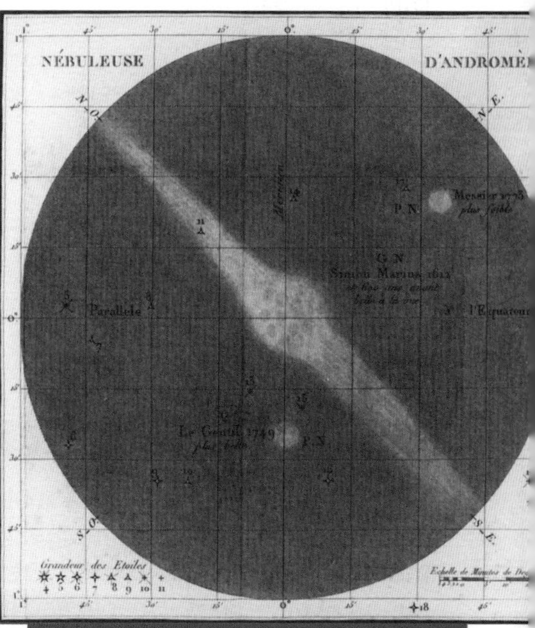

A The Andromeda nebula has been observed for over 1000 years. Charles Messier (1730–1817) published this drawing of it in 1807.

B The Andromeda Galaxy, as photographed by the Hubble Space Telescope.

We can also see a band of light across the sky on a clear, dark night. This is called the Milky Way. Through a telescope, it is clear that the band of light is caused by a band of stars concentrated in that part of the sky. In 1755, Immanuel Kant (1724–1804) speculated that the stars might be part of a huge rotating disc, and the different densities of stars in different directions were a result of looking out of the disc in different directions. He also speculated that some of the nebulae could be other **galaxies** like our own. Later astronomers studied the distributions of stars to try to estimate the size and shape of the galaxy.

?

1 What is a nebula?

2 a What evidence is there that the Sun is part of a disc-shaped galaxy?
  b How did the development of telescopes help to provide this evidence?

Harlow Shapley (1885–1972) mapped the distribution of globular clusters of stars, many of which contained Cepheid variables. His work led him to suggest that our galaxy is 300 000 light years across, with the Sun far from its centre. This large estimate of the size led Shapley to conclude that the Universe was one big galaxy and that all the nebulae were small gas clouds within the galaxy.

Heber Curtis (1872–1942) disagreed with this conclusion. His observations led him to suggest a much smaller size for our galaxy with the Sun near the centre. This meant that at least some of the nebulae were beyond our galaxy. The two men debated their ideas at a meeting of the National Academy of Sciences in Washington in 1920.

The debate was not resolved until 1924, when Edwin Hubble (1889–1953) discovered Cepheid variables in the Andromeda nebula. Hubble showed that the distance to these Cepheids was far greater even than Shapley's estimate of the size of the Milky Way, and therefore the nebula must be a separate galaxy. The Andromeda nebula is now called the Andromeda Galaxy.

C An image of distant galaxies taken using the Hubble Space Telescope.

?

3 Why was the presence of Cepheid variables in globular clusters important for:
a Shapley
b Hubble?

4 Why would Hubble have needed a series of photos of Andromeda to calculate its distance?

5 'Shapley was partly right, and Curtis was partly right.' Explain this statement.

6 Shapley overestimated the size of the Milky Way because he had not taken into account the effects of clouds of dust and gas in space. Explain how the presence of dust would have affected Shapley's distances.

Summary

Write a short magazine article outlining the different ideas of Shapley and Curtis, and explain how the debate was finally resolved.

P7.12

Distant galaxies

How do we work out how far away other galaxies are?

The parallax method can now be used to calculate the distances to stars up to around 650 parsecs away. Cepheid variables allow astronomers to calculate the distances to galaxies millions of parsecs away. However, the most distant galaxies observed to date are around 40 000 **megaparsecs** away (1 Mpc = 1 000 000 pc). Clearly, there must be some other ways of estimating these huge distances.

The spectra of stars and distant galaxies show absorption lines in patterns that depend on the elements present. In some cases the pattern of lines is shifted compared to similar stars. **Redshift** refers to lines moved towards the red (long wavelength) end of the spectrum. This shifting of the lines is due to the star or galaxy moving away from us.

In the 1920s, Edwin Hubble measured the redshifts in the data from many galaxies. Photo A shows some of his original data.

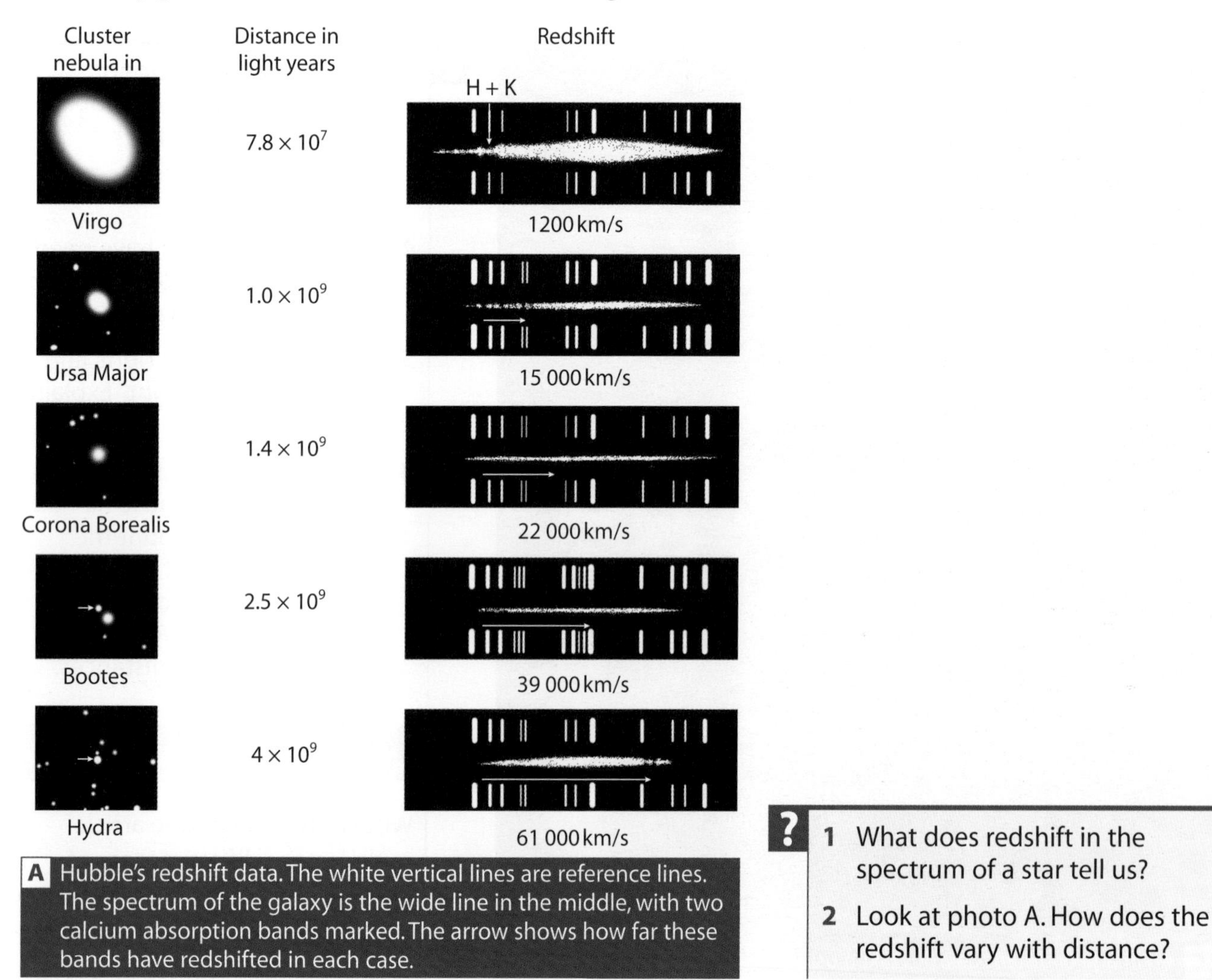

A Hubble's redshift data. The white vertical lines are reference lines. The spectrum of the galaxy is the wide line in the middle, with two calcium absorption bands marked. The arrow shows how far these bands have redshifted in each case.

?

1 What does redshift in the spectrum of a star tell us?

2 Look at photo A. How does the redshift vary with distance?

Hubble also used Cepheid variables to measure the distances to these galaxies, and found that the speed at which a galaxy was moving away depended on its distance. The further away the galaxy was, the faster it was moving away. All galaxies for which he could measure both distance and redshift followed this pattern. This movement is explained by the idea that the Universe as a whole is expanding, and means that we can use speeds from redshift measurements to work out the distances of galaxies that are too far away for other methods to be used.

E The speed of a galaxy moving away can be calculated using this equation:

speed = Hubble constant × distance
(km/s) (km/s/Mpc) (Mpc)

As more and more accurate instruments are developed, the distances to Cepheid variables in more and more distant galaxies can be measured. The value of the **Hubble constant** is around 70 km/s/Mpc. However, as new measurements are made, this value sometimes has to be adjusted.

The Whirlpool Galaxy is approximately 9.5 Mpc away. How fast is it moving away from us?

speed = Hubble constant × distance
= 70 km/s/Mpc × 9.5 Mpc
= 665 km/s

B The Whirlpool Galaxy.

?

3 Why did Hubble need to use information from Cepheid variables to work out the equation above?

4 The Sunflower Galaxy is 11.34 Mpc away, and the Southern Pinwheel Galaxy is 4.6 Mpc away.

a Without doing any calculations, how can you tell which galaxy is moving away from us more quickly?

b Calculate the speed of these two galaxies.

5 The Pinwheel Galaxy is moving away from us at approximately 580 km/s. How far away is it?

6 **a** Convert the distance data in photo A to Mpc. (*Hint*: you might need to look back at Topic P7.8.)

b Work out the Hubble constant for each speed and distance shown.

c How does the average Hubble constant worked out from these data compare with the value given on these pages? Suggest a reason for any difference.

Summary

Write a paragraph to answer the question at the top of page 136.

P7.13

Protostars

What happens when you compress a gas?

The particles in a gas are always moving and colliding with each other and with the walls of their container. The forces from all the collisions can be measured as **pressure**.

If you reduce the volume of a gas, the molecules have less space to move around in. They are more likely to collide with each other and with the walls of the container, so the pressure increases.

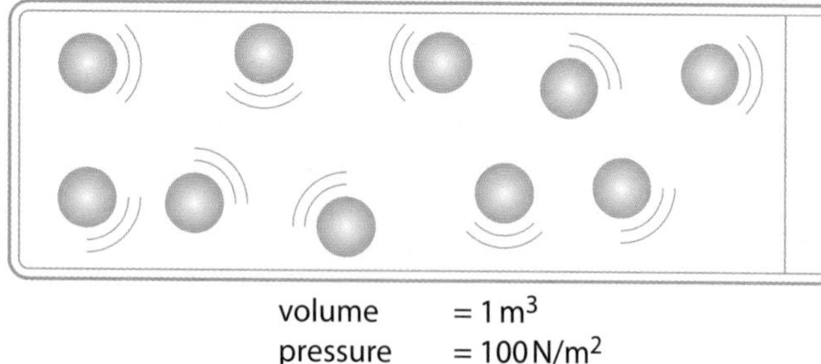

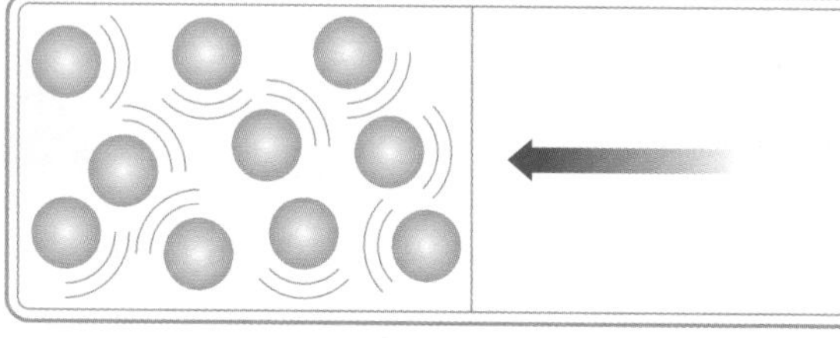

A Halving the volume doubles the pressure.

If you increase the temperature of the gas, the molecules have more energy. They move faster and collide with each other more often. As there are more collisions, the pressure is higher (as long as the volume remains constant).

As you reduce the temperature of a gas, the molecules move more slowly because they have less energy. When you have cooled it enough, the gas condenses into a liquid. As you cool the liquid further, the molecules move even more slowly and the liquid then freezes into a solid. The molecules are not moving around in the solid, but they *are* vibrating.

?

1 What causes pressure in a gas?

2 What will happen if you reduce the temperature of a gas while keeping its volume constant?

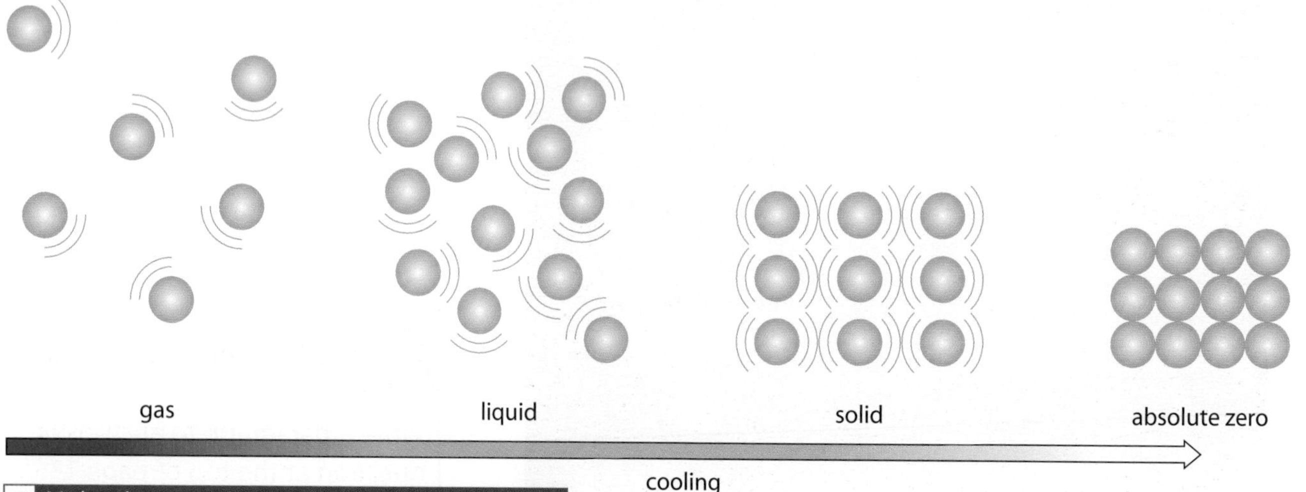

B Molecules move less and less as they get cooler.

As the temperature of a solid decreases, the molecules vibrate less and less until they stop vibrating completely. This happens at –273 °C for all substances. This temperature is known as **absolute zero**.

The **Kelvin temperature scale** takes absolute zero as its starting point. The unit is the kelvin (symbol K) and is the same size as a degree Celsius. 0 K is –273 °C.

E

temperature (in kelvin) = temperature (in degrees Celsius) + 273

temperature (in degrees Celsius) = temperature (in kelvin) – 273

Human body temperature is 37 °C.
In kelvin, this is 37 + 273 = 310 K.

Helium boils at 4 K.
In degrees Celsius, this is 4 – 273 = –269 °C.

?

3 Why do the molecules move more slowly when you cool a gas?

4 What is absolute zero?

5 What are the freezing and boiling points of water on the Kelvin scale?

6 What are the following temperatures in °C?
a 473 K
b 298 K

Stars form from clouds of gas. Gravity pulls the molecules in the gas closer and closer together. As the gas is compressed, the temperature increases. As the cloud of gas gets denser and hotter it exerts a bigger force of gravity, which pulls in even more gas. A cloud of gas like this is called a **protostar**, which is a star before it has started to produce its own energy. As it pulls in more gases, it gets denser, which continues to increase its temperature.

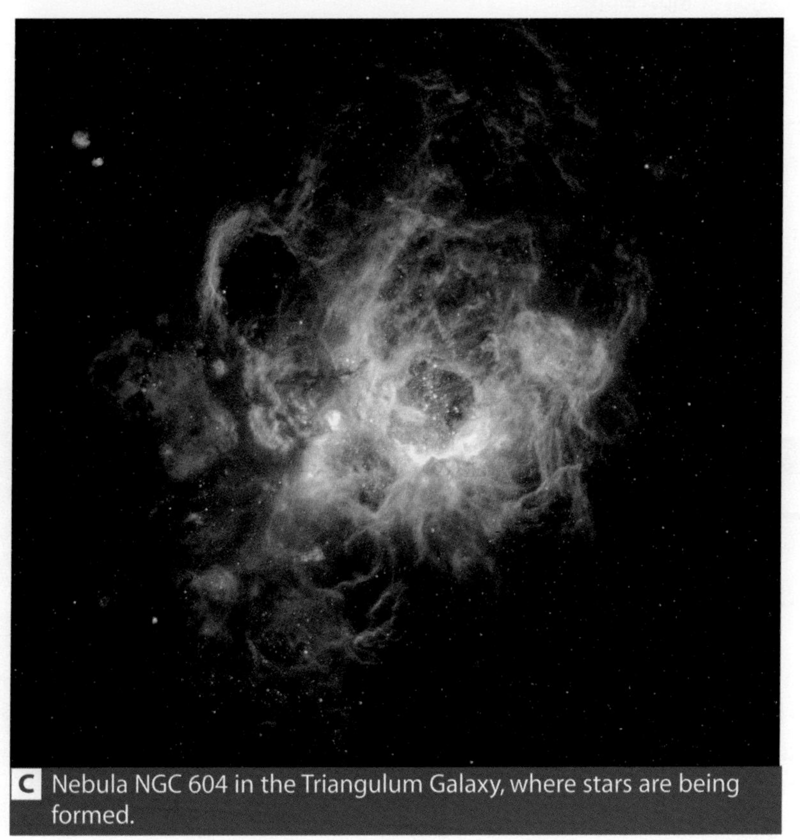

C Nebula NGC 604 in the Triangulum Galaxy, where stars are being formed.

?

7 **a** What is a protostar?
b How is a protostar formed?

8 Sketch and label a graph showing the volume of a gas against its temperature. Include absolute zero on your graph.

Summary

Write an information leaflet or poster to show how a protostar is formed. Explain what happens in terms of pressure, temperature and volume.

Nuclear energy

How do stars produce energy?

The electron was discovered in 1897, but protons and neutrons had not yet been discovered. At the beginning of the 20th century it was thought that an atom was a sphere of positive charge that contained electrons.

In 1909, Hans Geiger (1882–1945), Ernest Marsden (1889–1970) and Ernest Rutherford (1871–1937) fired alpha particles at a piece of gold foil. They expected that some of the alpha particles would be deflected by a few degrees as they passed through the gold atoms. They were surprised to find that a few alpha particles were deflected by more than 90 degrees, which meant that they must have been deflected by a large force.

They concluded that atoms were not solid at all but consisted of a small and very dense **nucleus** that has a positive charge, with electrons orbiting it. Most of the alpha particles had passed straight through the gold atoms, but some of them were repelled by the positive charge of the nucleus.

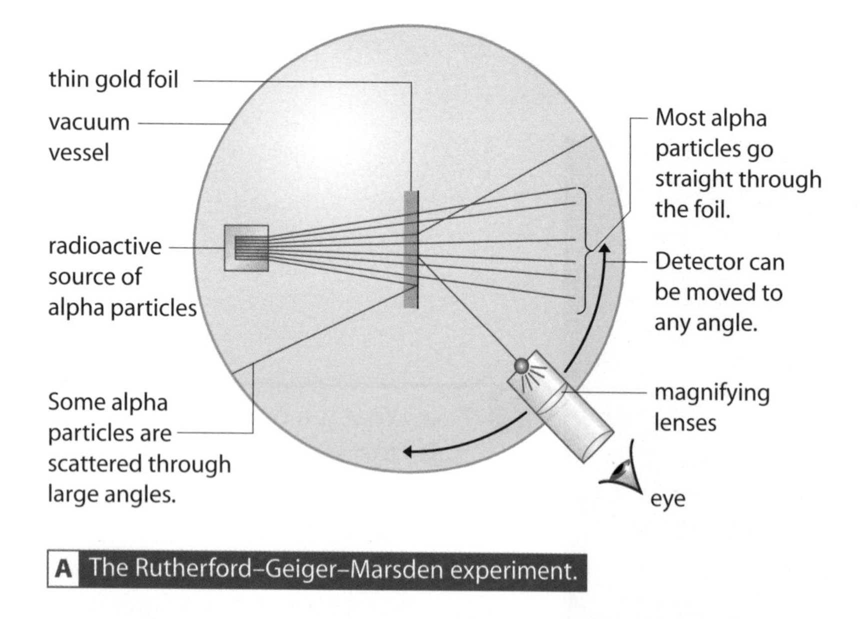

A The Rutherford–Geiger–Marsden experiment.

?

1 What happened when alpha particles were fired at gold foil?

2 Why were the alpha particles repelled by the nucleus?

We now know that the nucleus is made up of protons and neutrons. Protons were discovered in 1918 and neutrons in 1932. Protons and neutrons are very small and the nucleus of an atom is small and massive. Most of the rest of the atom is empty space, except for electrons orbiting the nucleus.

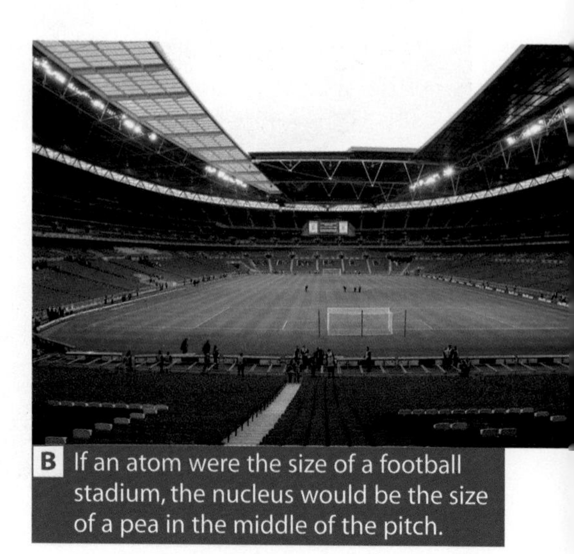

B If an atom were the size of a football stadium, the nucleus would be the size of a pea in the middle of the pitch.

You might expect the positively charged protons in the nucleus to repel each other. However, the protons and neutrons in the nucleus are held together by a **strong nuclear force**, which is much stronger than the force caused by the positive charges.

The nucleus of a hydrogen atom is a single proton. If you try to push two protons together, they will repel each other. If you keep on increasing the force so that the protons are forced together, eventually they will join to make a single, larger nucleus. When this happens, energy is released. The process is known as **fusion**.

3 Why don't protons repel each other in the nucleus?

4 What happens when you force two protons together?

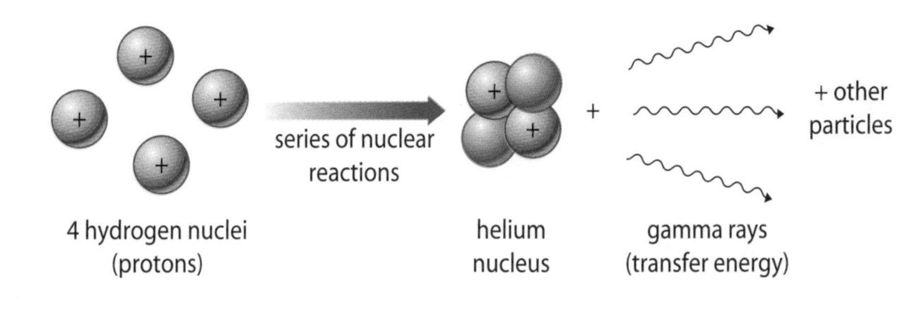

C Hydrogen is converted to helium inside stars.

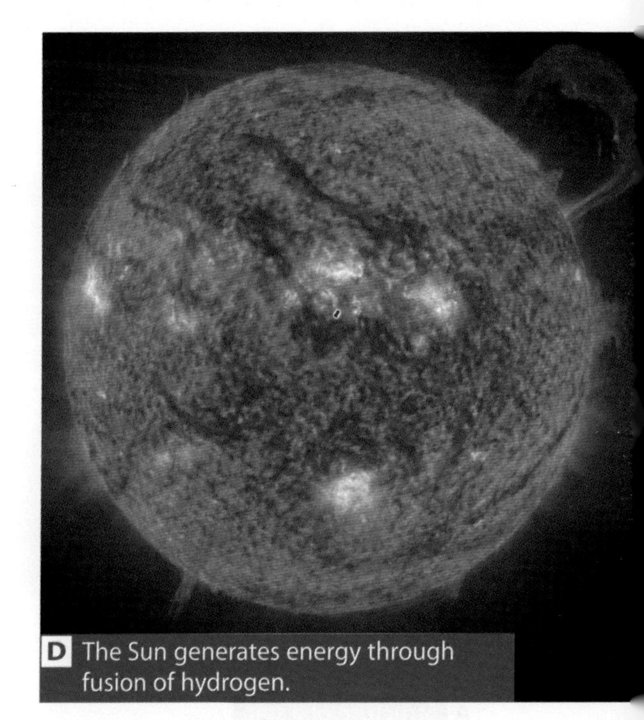

D The Sun generates energy through fusion of hydrogen.

At the beginning of the 20th century, scientists were puzzled about the source of energy in the Sun. It had been thought that coal was the source of the Sun's energy, but calculations showed that burning coal would not provide enough energy.

Einstein's famous equation linking mass and energy provided a clue. When two atoms fuse, some of the mass is transferred to energy. It is only a small mass but a lot of energy. For example, when 1 tonne of hydrogen fuses to make helium, it releases the same amount of energy as burning about 35 000 million tonnes of coal!

?

5 Why did scientists at the beginning of the 20th century need a new theory to explain the source of the Sun's energy?

6 How does a star produce energy?

7 Explain the difference between nuclear fusion and nuclear fission.

Summary

Write a short script for a radio programme outlining the discovery made by Geiger, Marsden and Rutherford and how it changed our ideas about the atom. Include ideas about how nuclei can fuse.

P7.15 Red giants

What will happen when the Sun runs out of fuel?

Stars like the Sun emit energy that is released when hydrogen nuclei fuse to form helium. There are several stages in the process that involve the formation of other isotopes of hydrogen (deuterium and tritium) before helium is formed.

Fusion takes place in the central **core** of a star. The energy released radiates away from the core. In the outer layers of the star, the energy is taken to the surface by **convection**. At the surface of the star, the **photosphere**, the energy is radiated into space. This energy causes an outward radiation pressure. This balances the gravitational force that is trying to collapse the star.

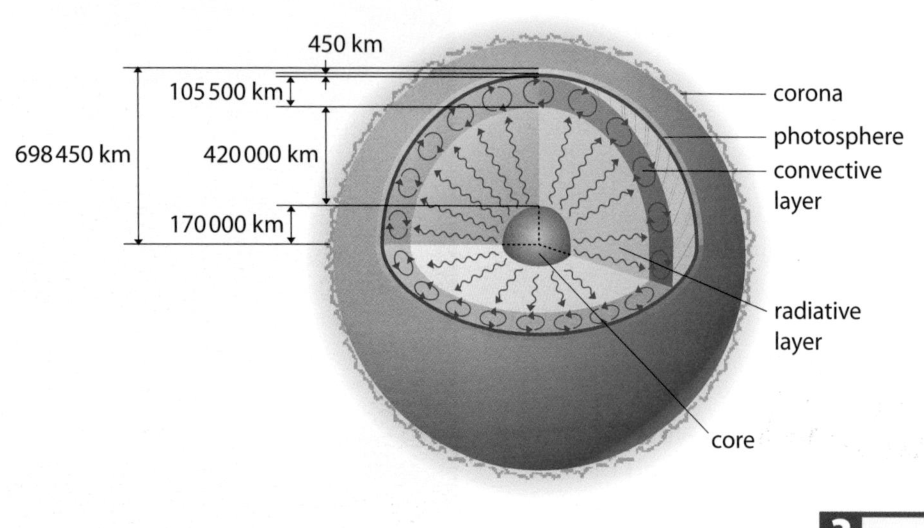

A Structure of a star.

? 1 a How does a star produce its energy?
b Where does this happen?

As the star uses up hydrogen, helium accumulates in the core. Eventually, all of the hydrogen in the core of the star has been used up, and the core is made up of helium.

The core starts to collapse under gravity. The pressure increases and so does the temperature. When the core is hot and dense enough, helium nuclei start fusing to form bigger nuclei such as carbon, nitrogen and oxygen.

As this happens, the outer layers expand so that the star increases between 70 and 100 times in size. Since the surface of the star is cooler, it looks red. When the Sun becomes a **red giant** it will expand and engulf Mercury and Venus, and come much closer to the Earth.

B The Sun will become a red giant in about 4 billion years' time.

When all the helium has been used up, the core shrinks again. When the star has a mass similar to the Sun's mass, the pressure and temperature in the core are too low for any more fusion to take place. The star becomes a **white dwarf**. As this happens, the outer layers of the star expand and matter is ejected from the star to form a **planetary nebula**.

?

2 **a** Why does a star start to collapse when all its hydrogen has been used up?
b Why do the pressure and temperature in the core increase when this happens?

3 Why does a star that is cooler than the Sun look red? (*Hint*: you may need to look back at Topic P7.9.)

4 What elements are created by the fusion of helium?

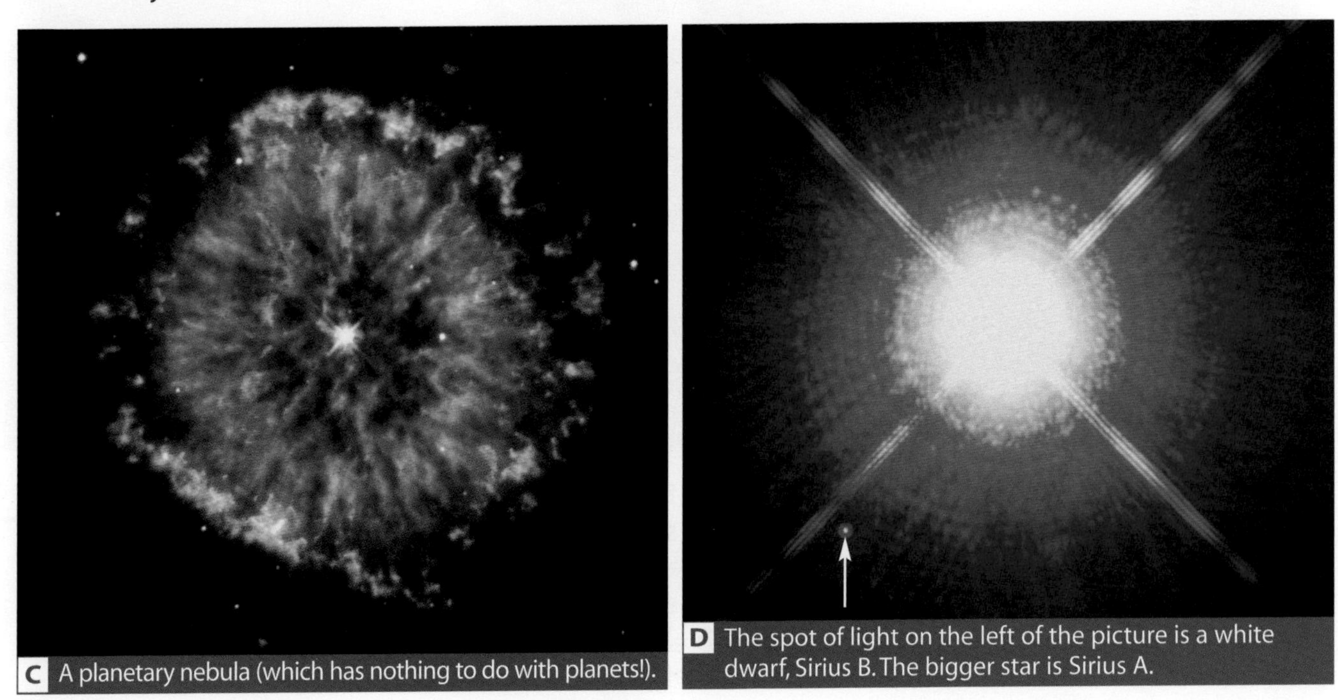

C A planetary nebula (which has nothing to do with planets!).

D The spot of light on the left of the picture is a white dwarf, Sirius B. The bigger star is Sirius A.

White dwarfs are very small and very dense. They have a mass similar to that of the Sun but are a similar size to the Earth. They are not very bright and gradually cool down.

?

5 Why does carbon not fuse to form larger atoms in a red giant?

6 How much denser is a white dwarf than the Sun? Assume that the radius of the white dwarf is about 6000 km and the radius of the Sun is 700 000 km.

Summary

Write a paragraph to describe what happens to stars like the Sun when they run out of hydrogen in the core.

P7.16 Red supergiants

What happens when stars bigger than the Sun run out of fuel?

When a star like the Sun runs out of hydrogen in its core, it forms a red giant. Helium is fused to form larger atoms such as carbon, nitrogen and oxygen. Stars that have a much higher mass than the Sun also expand when they run out of hydrogen fuel, but they become **red supergiants**, which are much bigger than red giants.

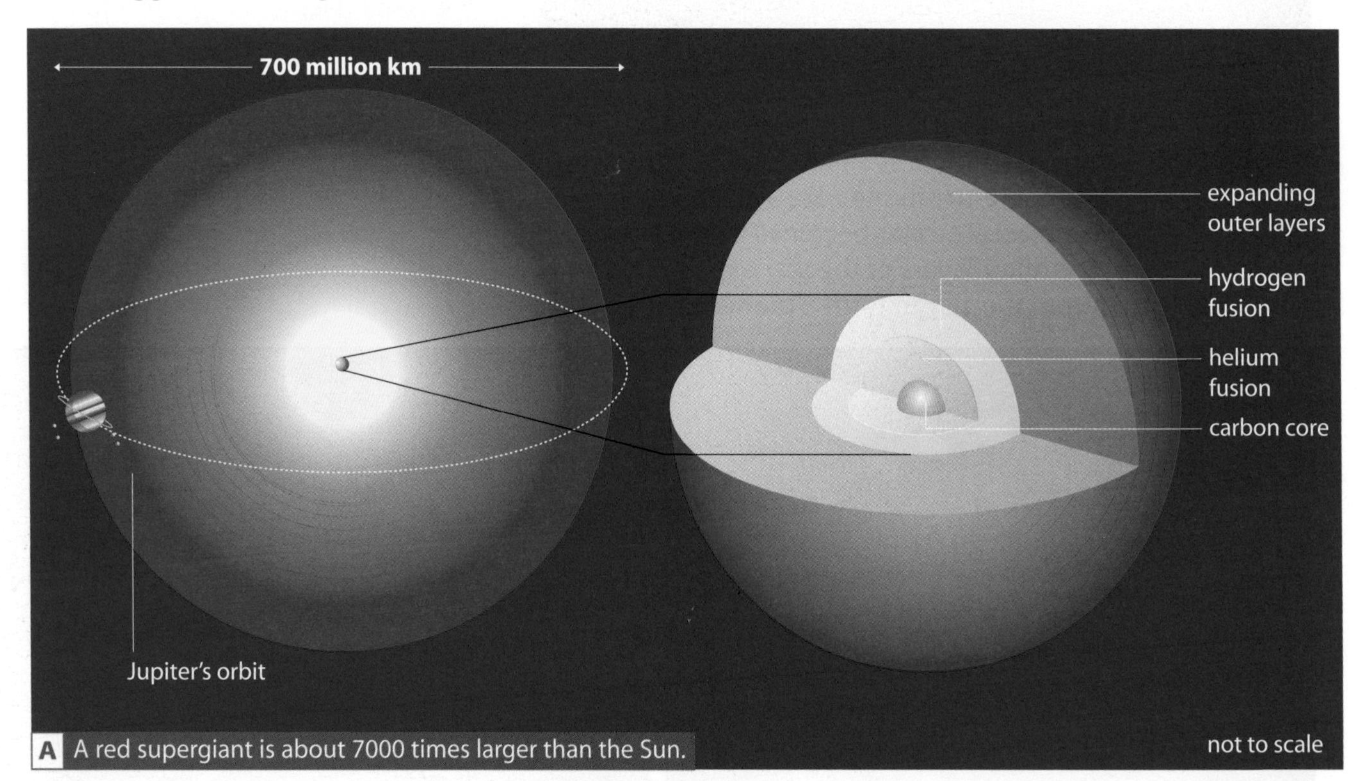

A A red supergiant is about 7000 times larger than the Sun.

When the helium in the core has run out and been fused to make carbon, oxygen and nitrogen, the core of the star contracts again. The pressure and temperature increase enough for further fusion to take place. Other elements up to iron in the Periodic Table are created by fusion. This happens in stars whose core mass is more than 1.4 times that of the Sun.

When elements bigger than iron are created by fusion, energy is taken in to make the fusion happen. No energy is given out. This means that nuclei bigger than iron cannot be formed in the core of a red supergiant.

There can be many different fusion reactions going on in layers in the core. At the centre of the core, iron is being created, whereas at the outside of the core, hydrogen is still being fused to make helium. Other fusion reactions are taking place in layers in between. The core of a red supergiant is surrounded by a very large layer of hydrogen.

?

1 a What are the similarities between red giants and red supergiants?

b What are the differences?

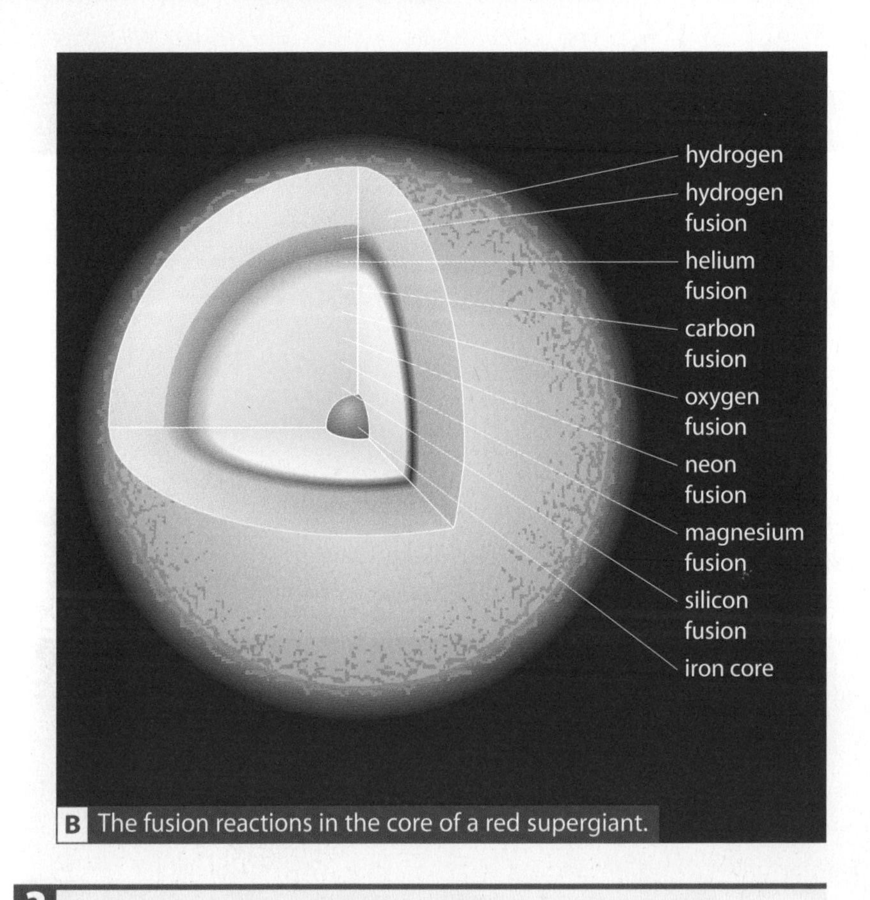

B The fusion reactions in the core of a red supergiant.

?

2 Why are no atoms bigger than iron created in fusion?

3 In a red supergiant, what is being created by fusion at the:
a centre of the core
b outer edge of the core?

When the core of the star is mostly iron, the iron core then collapses and heats up. The atoms become so tightly packed that they turn into neutrons. This core, which is about the size of the Earth, collapses even further, to about 10 km across, in about 1 second. The outer layers are pulled in by gravity and heat up to billions of degrees Celsius, and then explode in a **supernova**. Many supernova explosions have been observed. They usually last only a few weeks.

The core that is left is a **neutron star**. If the mass of the original star was more than five times the mass of the Sun, the core collapses even further and becomes a **black hole**.

?

4 What happens to a red supergiant when the centre of the core is made of iron?

5 What is a supernova?

6 Will the Sun explode in a supernova? Explain your answer.

7 Find out the density of a neutron star, and compare it to the density of lead.

C The Crab Nebula is the remains of a star that exploded in a supernova about 1000 years ago. People on Earth could see it without a telescope.

Summary

Draw a flow chart to show what happens to three stars: Star A has a similar mass to the Sun, Star B is twice the mass of the Sun, and Star C is four times the mass of the Sun.

P7.17 Observing the stars

Where are large telescopes built?

You cannot see stars on a cloudy night, but even on a clear night you may not be able to see many stars. In many parts of the UK, light from streetlights, buildings and other places is far brighter than the faint light from the stars. You cannot see the stars because of **light pollution**. Moisture in the atmosphere can also refract the light from the stars and blur the images.

To see the stars more clearly, you need to go to a remote place where there is no light pollution. You also want somewhere that is more likely to have clear skies, and with little dust or moisture in the atmosphere. The ideal place for a telescope is at the top of a high mountain in a much drier region than the UK.

A Light pollution from the Earth.

B Telescopes on the top of Mauna Kea, Hawaii.

?

1. Why can't you see many stars from many places in the UK?
2. What conditions are best for observing stars?

The summit of Mauna Kea on the island of Hawaii is at 4205 metres above sea level. The climate is fairly dry and the skies are usually clear. Several telescopes have been built there. Another good place for telescopes is the Atacama desert in Chile.

Astronomers can work at the site of the telescope. They can move the telescope so that it is pointing at the part of the sky they want to observe and then use instruments to record images. In a large telescope, this is done by entering the coordinates of the object they want to look at into a computer. The computer controls the motors that move the telescope so that it points to the correct place in the sky.

C Working at a telescope.

Telescopes are very often in remote places, so astronomers may not go to the telescope to carry out their observations. Instead, they give detailed instructions to a technician at the telescope who carries out their observations for them and then sends them the results.

Some telescopes can be controlled remotely over the internet using a computer. In this way, the telescope can be not only pointed at the correct spot in the sky but also programmed to track an object accurately across the sky so that long-exposure photographs can be taken. This means that people do not have to work in remote places.

?

5 What are the advantages and disadvantages of working remotely from a telescope?

6 What are the advantages of using a computer to control a telescope?

7 Jodrell Bank Observatory is located not far from the city lights of Manchester. Find out why light pollution does not matter for this observatory.

?

3 What problems can be avoided by putting telescopes at the top of Mauna Kea?

4 What do you think the conditions in the Atacama desert are like?

Summary

Imagine you are an astronomer. Write a paragraph to describe what your work involves.

P7.18 Telescopes in space

Why are telescopes used in space?

Sometimes it is very difficult to observe stars from the surface of the Earth. You need clear skies, dry air and no light pollution. Even when the atmospheric conditions are good, the Earth's atmosphere still absorbs about 10% of visible light. Other parts of the electromagnetic spectrum are absorbed completely by the upper atmosphere. The atmosphere can also refract light and reduce the sharpness of images.

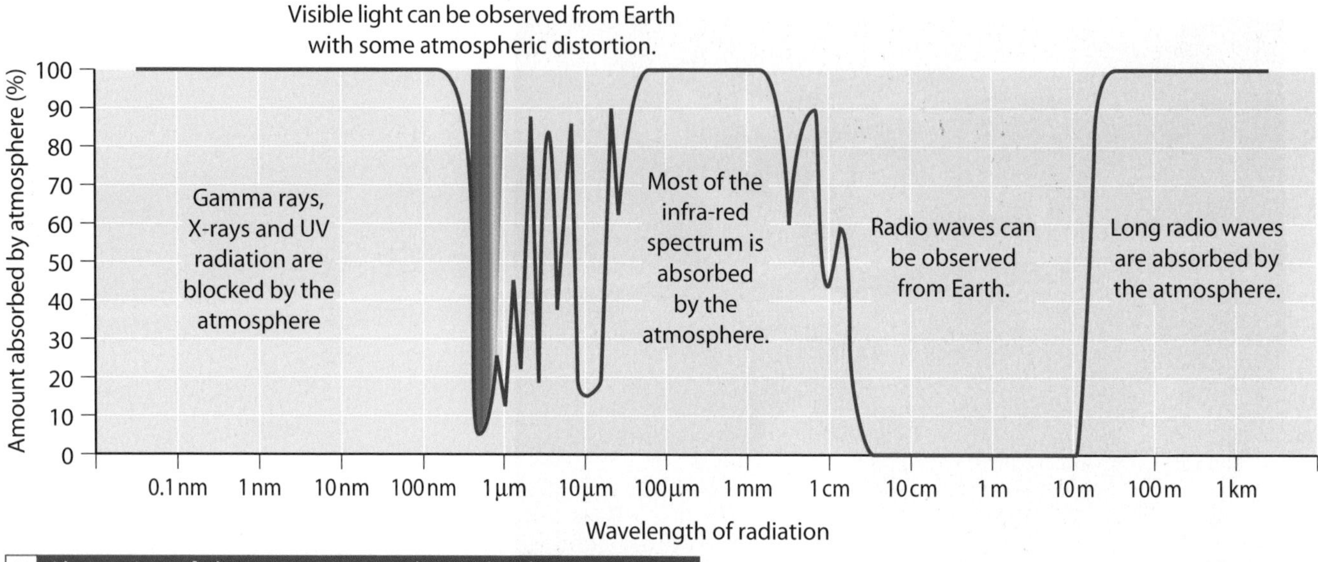

A Absorption of electromagnetic radiation by the atmosphere.

One way of improving the quality of observations is to have the telescope outside the Earth's atmosphere. The Hubble Space Telescope (HST) was launched into orbit in April 1990. It has been very successful, allowing astronomers to make much more detailed observations than they could before.

Initially, the HST was a disappointment. Astronomers found that the images were not as sharp as they had hoped. When something is wrong with a car, you can take it to a mechanic who will examine the car to find out what is wrong with it. The astronomers could not do this. Instead, they had to analyse the observations from the telescope to try to work out the source of the error.

Eventually, they found that the mirror had been made very accurately, but the specification was wrong. The mirror was too flat around the edges, so the rays of light were not all being focused in the same place.

This was corrected by using two new mirrors that would compensate for the incorrect shape of the original mirror. The mirrors were fitted by astronauts from the Space Shuttle in December 1993.

?

1. What are the disadvantages of observing stars from the surface of the Earth?
2. What are the advantages of observing stars from space?
3. What can be observed from space that cannot be observed from the surface of the Earth?
4. Why was it difficult to find out what was wrong with the HST?
5. What problems do you think astronauts might have had when fitting the new mirrors?

Maintaining and repairing the HST is very expensive. Each mission of the Space Shuttle costs many millions of pounds. The original cost of the telescope was approximately £100 million. A proposed repair mission for 2008 was estimated to cost nearly £500 million.

If something on the HST stops working, astronomers have to wait until a repair mission can be fitted into the Space Shuttle's programme. If something goes wrong with the Space Shuttle, they may have to wait years for a repair.

B Astronauts from the Space Shuttle servicing the HST.

C The James Webb Space Telescope is due to be launched in 2013 to replace the HST.

?

6 What are the disadvantages of using a telescope in space?

7 The HST has a door that can close over the telescope. Suggest why this is necessary.

Summary

Draw up a table to show the advantages and disadvantages of using telescopes in space.

P7.19 New telescopes

How are new telescopes planned and built?

Astronomers want to observe further and further into the Universe and make more detailed observations of stars and galaxies. They would like to look for planets like the Earth, but they need very powerful instruments to do so.

A Astronomers need very powerful telescopes like this one to observe different galaxies.

To look further, telescopes need to be bigger. This means that the mirrors need to be bigger. The largest mirrors in telescopes are currently about 10 metres across. There are plans for new telescopes with much bigger mirrors – up to 40 metres across.

Telescopes are high-precision instruments that are expensive to make and can also be expensive to run. To justify these high costs, they need to be used as much as possible. To ensure this, many organisations are involved in the building and running of telescopes, and as a result, many more astronomers have access to them and can make observations.

This has the advantage that the costs are shared between many organisations, often from several countries. It also means that there is a much bigger pool of expertise available for the organisations running the telescope.

In 2005, the Southern African Large Telescope (SALT) was completed. Organisations from South Africa, the USA, Germany, Poland and New Zealand are involved. The organisations in the UK include a consortium of about six universities. SALT has an 11-metre primary mirror and is located in the semi-desert in the Karoo, South Africa. It can detect objects as faint as a candle flame on the Moon!

?

1. Why do astronomers need better telescopes?
2. Why do you think that astronomers want to look for planets?

B The Southern African Large Telescope.

?

3 What are the advantages of collaboration when building a new telescope?

4 How many countries are involved in SALT?

5 Give two reasons why it would be difficult to spot a planet orbiting another star.

When a new telescope is built, there are non-astronomical factors that have to be taken into account. For example, many telescopes are on mountain tops. The top of a mountain can be a very fragile environment that can easily be damaged and take many years to recover.

Telescopes are often built in very remote places with few, if any, facilities. Astronomers need somewhere to stay that is not too far away. The telescopes on Mauna Kea, Hawaii, are at 4200 metres above sea level. Living quarters were needed nearby, but they were built part-way down the mountain at about 2900 metres, to minimise the effects of altitude sickness.

C Accommodation had to be built on Mauna Kea for astronomers.

?

6 What are the problems with choosing a site at the top of a mountain?

7 Why was the astronomers' accommodation on Mauna Kea built part-way down the mountain?

8 Why is the primary mirror made up of hexagonal segments in most large telescopes?

Summary

You are asked to write a project proposal for a new telescope. Write a brief report to summarise the factors that you need to take into account.

Revision questions

1 Astronomers can record the position of a star using angles. Diagrams A and B show two different ways of doing this.
   a Which way of referring to the position of a star is used by astronomers?
   b Explain why astronomers find this the most useful way.

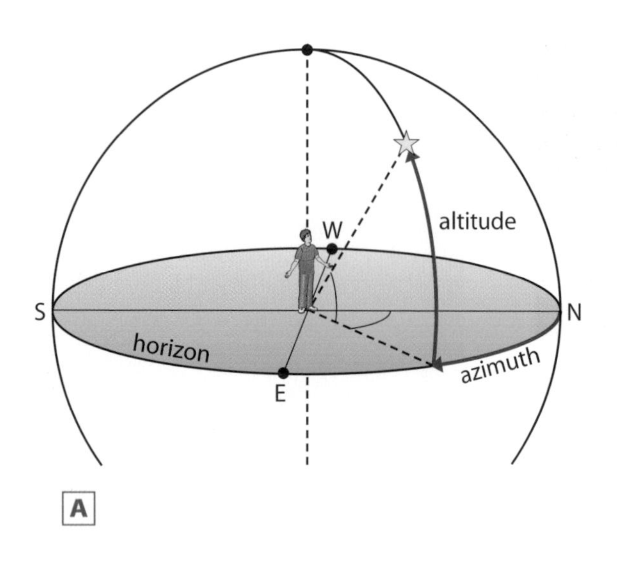

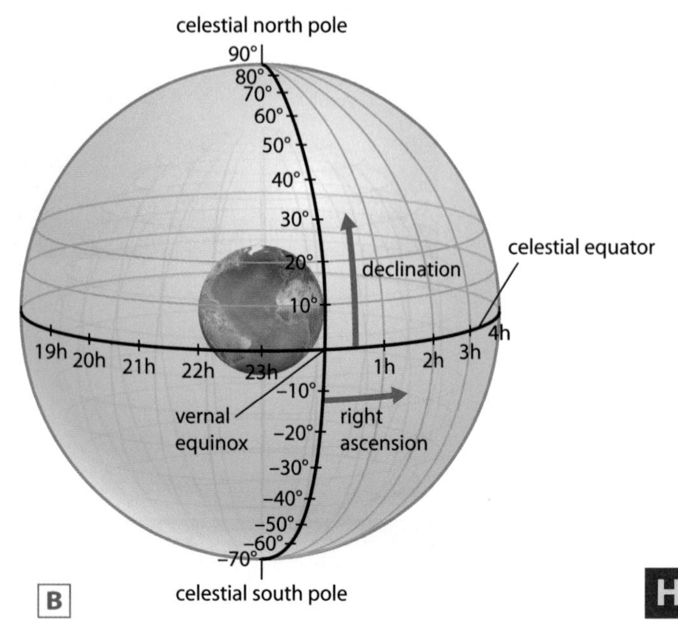

2 The stars appear to travel across the sky once in 23 hours 56 minutes. This time is known as a sidereal day.

The Moon appears to travel across the sky once every 24 hours 49 minutes.

   a Explain why the Moon appears to take more than a sidereal day to travel across the sky once. Use a diagram to help you to explain.
   b Explain why the Moon does not always appear to have the same shape. Use a diagram to illustrate your answer.

3 a If you live in the UK, you can see the star Antares in June but not in December. Use a diagram to help you explain this statement.
   b The star Acrux is never visible from the UK. Explain why there are some stars that can never be seen from the UK.

4 Lens A has a focal length of 4 cm.
   a Draw a diagram to show how this lens will bring parallel rays of light to a focus. Use the correct distances on your diagram.
   b Lens B is made from the same material as lens A but its surfaces are more curved. How will it focus light differently from lens A?

5 The diagram shows a telescope.

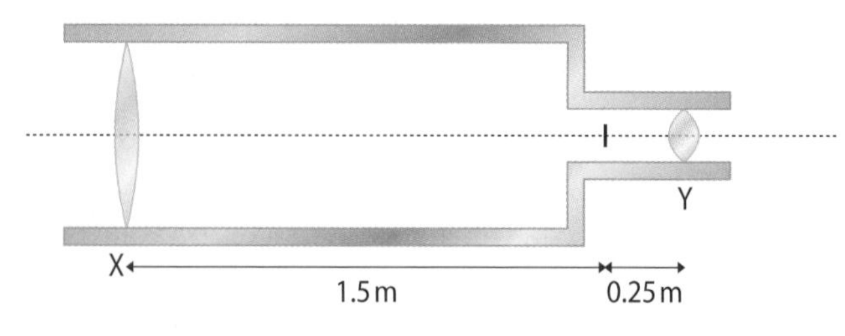

   a Which lens is the eyepiece? Explain your answer.
   b What is the power of lens X?
   H c What is the angular magnification of this telescope?

6 **a** What is a Cepheid variable? Explain in as much detail as you can.
b Explain the difference between the terms 'apparent brightness' and 'intrinsic brightness' of a star.
c Describe how the distance to a Cepheid variable can be estimated.

7 A nebula is a cloud of dust or gas in space, but originally the word was used to describe any fuzzy object seen through a telescope.
a The astronomers Curtis and Shapley had two different explanations for what nebulae were. What were their two ideas?
b How did Edwin Hubble's observations of Cepheid variables in one nebula help to determine which of these ideas was correct?

8 The light from most galaxies shows redshift.
a What is redshift?
b What does the amount of redshift in the light from a galaxy tell us about that galaxy?

9 Edwin Hubble worked out this equation that relates the distance of a galaxy to the speed at which it is moving away from us:

speed of recession (km/s) = Hubble constant (km/s/Mpc) × distance (Mpc)

The Hubble constant is approximately 70 km/s/Mpc.

a What **two** things did Hubble need to measure to help him work out this equation?
b What unit does 'Mpc' represent?
c Galaxy NGC 1300 is approximately 20 Mpc away. How fast would you expect it to be moving away from us?
H **d** Galaxy NGC 4414 is moving away from us at approximately 700 km/s. How far away would you expect this galaxy to be?

10 **a** A star contains different layers. Draw a diagram of a star, and label the three layers.
b Briefly describe what happens in each layer of a star similar to the Sun.

11 **a** Draw a diagram of an atom, and label the nucleus, protons, neutrons and electrons.
b The results of the Rutherford–Geiger–Marsden experiment showed that an atom is mostly empty space. Describe this experiment, and explain how the results led to this conclusion.
c Protons are positively charged, so you would normally expect them to repel each other. How can protons stay together in an atom?

Pre-release question

You will get a pre-release paper before the P7 exam. The text and questions here are to show you the kinds of questions you might have to answer, and to give you some hints about how to use the pre-release passage to help you to revise for the exam.

Pulsars

In 1967 Jocelyn Bell was a research student using a radio telescope to study different galaxies. When she examined the print-outs from the telescope, she noticed that there was a regular, pulsing signal coming from one part of the sky. Initially Bell and Anthony Hewish, her supervisor, wondered if these signals were signs of an extra-terrestrial civilisation, and jokingly referred to the signals as LGM1 (for 'Little Green Men 1').

The data were checked carefully to make sure the signal was not due to interference from radio signals on Earth. Once other, regular signals like this were discovered, it was confirmed that they had some natural cause, and they were named pulsars. Pulsars have also been observed with telescopes that look at other wavelengths in the electromagnetic spectrum.

Further investigations of pulsars have found that they are very small, dense stars that were probably formed in supernova explosions. The current theory is that pulsars are rotating neutron stars which emit beams of radiation from their magnetic poles. The star itself is not pulsing, but it is spinning rapidly. Radio telescopes on Earth detect the radiation only when the beam is pointing towards the Earth.

There is still a lot to be learnt about pulsars, and astronomers from many countries are still studying them.

A

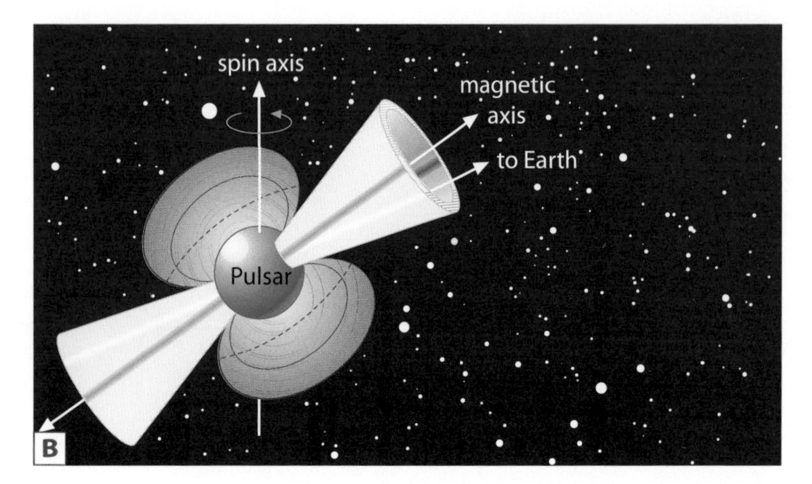

B

When you get your pre-release paper, read it carefully and then think about which areas of science it covers. Go back to the module where this was covered and revise any of the content that looks relevant to the passage. The modules you need may be in the GCSE Science or the GCSE Additional Science book. Also think about the 'Ideas about Science' that might be relevant to the text (see the list on page 113).

The passage above is about pulsars, which are a kind of neutron star. Neutron stars are covered in Module P7 in this book, but the life cycles of stars were also covered in Module P1. The text above also mentions the possibility of signals from extra-terrestrial intelligences, which was also looked at briefly in Module P1. It also mentions different parts of the electromagnetic spectrum, which was covered in Modules P2 and P6.

Some of the 'Ideas about Science' that could be asked about include how scientists develop explanations for observations, how scientific findings are reported and checked, and how new theories, ideas or discoveries are accepted or rejected by the scientific community. These ideas were covered in P1 in connection with ideas about continental drift and plate tectonics.

If the passage above was given to you as your pre-release paper, it would be a very good idea to revise what you learned in P1, and also to take a look at P2 and P6 before the exam. This doesn't mean that you *will* get questions on these modules, just that you *might*.

The questions below are similar to the type of question you may get in the exam. You will not see the questions before the exam, only the text.

1 Radio waves are part of the electromagnetic spectrum with long wavelengths.
Which part of the electromagnetic spectrum has:
a the lowest photon energies [1 mark]
b the shortest wavelengths? [1 mark]

2 **a** Suggest why a pulsing radio signal might have been due to an extra-terrestrial civilisation. [1 mark]
b Why was it decided that the signals were not due to 'Little Green Men'? [2 marks]

3 Suggest what the mass of a pulsar would be, in comparison to the mass of our Sun. Explain your answer. [2 marks]

4 'Life on Earth could not exist if there had never been supernova explosions.'
Explain this statement, with reference to nuclear reactions inside stars.
One mark is for a clear, ordered answer. [3 marks + 1 mark]

5 Describe **two** different ways in which other astronomers could have found out about the discoveries made by Bell and Hewish. [2 marks]

6 Give **two** advantages of astronomers around the world collaborating to study pulsars. [2 marks]

Coursework

Practical Data Analysis and Practical Investigation

One of your pieces of coursework is to plan, carry out, analyse and evaluate a complete practical investigation (the Practical Investigation). One of the other pieces of coursework is to analyse and evaluate data that you have gathered or that has been given to you (Practical Data Analysis). The Practical Data Analysis is marked in exactly the same way as the 'Interpreting' and 'Evaluation' parts of the Practical Investigation.

Your Practical Data Analysis and your Practical Investigation coursework will be marked by your teacher, and checked by the examiners.

Ask your teacher to let you have a copy of the *Data analysis checklist* or the *Practical investigation checklist*. These show what you have to do to get good marks for your data analysis or investigation.

These pages outline the kinds of things you need to do or to think about when carrying out your investigation. There are also some Skills Sheets which your teacher may give you that contain more detail.

Strand S: Strategy

The first set of marks for your Practical Investigation is for planning your investigation. The question you choose to investigate can be linked to any module in the course. For example, the investigation for your GCSE Biology can be taken from any of the work in Modules B1 to B7.

Your teacher may suggest the topic for your investigation, or you may be allowed to choose your own subject. Not all topics that you could investigate will allow you to get top marks in all the strands, so do discuss your choice with your teacher before you spend too much time planning it.

In this strand you get the highest marks if:

- the investigation requires very accurate or precise measurements
- you plan your method yourself, and choose the apparatus you will use, rather than relying on instructions on a worksheet or from your teacher
- you choose the investigation yourself and carry it out independently – this does not mean you are not allowed to ask your teacher for a little help, but you should make most of the decisions yourself.

Planning an investigation can give you more hints for this strand.

Strand C: Collecting data

These marks are for how well you carry out your investigation, including how fair your test is.

You get the highest marks if:

- you have reviewed all the factors that might affect the outcome of your investigation, and described how you have controlled all the factors except the one you are investigating
- you have collected data over a big enough range, and repeated measurements to check for reliability
- you have carried out preliminary tests to establish the ranges for your factors, or to check that your apparatus or techniques are suitable
- you have used the apparatus carefully and accurately.

The skills sheets *Presenting data*, *Line graphs*, and *Pie charts* can help with this strand.

Strand I: Interpreting data

These marks are for how well you display and analyse your results and how well you present your conclusions. They apply to the Practical Data Analysis part of your coursework, as well as to the Practical Investigation.

You get the highest marks if you:
- display your results as line graphs or scatter plots, with lines of best fit
- use error bars to indicate the spread of data
- use calculations to help you to interpret your data, such as calculating the gradient of graphs, or using statistical methods
- identify the relationships between your factor (such as one factor being proportional to another)
- use scientific knowledge to explain your conclusion.

Strand E: Evaluation

These marks are for saying how well you think you carried out the investigation, and how you would modify the investigation if you had time to do it again. They apply to the Practical Data Analysis part of your coursework, as well as to the Practical Investigation.

You get the highest marks if you:
- describe how you could improve the apparatus or techniques you used, and give reasons why your suggestions would be an improvement
- consider how reliable your evidence is, by referring to the pattern of your results or the amount of scatter
- account for any anomalous measurements
- discuss how confident you are in your conclusions by referring to the apparatus or techniques used and the reliability of the data
- suggest what further data could help to make the conclusions more secure.

Skills sheet X *Evaluating investigations* can help you with this strand.

Strand P: Presentation

These marks are for how clearly and effectively you report what you have done.

You get the highest marks if you:
- clearly describe what you were trying to find out and clearly present details of your practical procedures
- include all the data you collected, including the actual data as well as averages or the results of any other calculations you have carried out
- labelling of tables is clear and correct, and all data is given to a suitable level of accuracy
- use correct scientific vocabulary
- spelling, punctuation and grammar are almost faultless.

Case Study

The case study is part of your coursework, and you will need to do three altogether (one for Biology, one for Chemistry and one for Physics). You are asked to present a report based on a detailed study of a chosen topic. The case study is designed to help you to find information in the media, to decide how scientifically accurate it is, and to make decisions based on the information you have found.

Your teacher will mark your case study, and it will be checked by the examiners. Ask your teacher for a copy of the *Case Study checklist*, which shows you what you have to do to obtain top marks.

Choosing a topic

The topic for your case study should:
- be based on science that is in the news
- involve a question that you can answer as part of your report
- be based on any area of science that you have covered in the course. For example, your physics case study can be based on any of the science in modules P1 to P7.

The question you try to answer in your case study could be:
- evaluating a claim where scientific knowledge is not certain (for example, 'Is there life in other places in the Solar System?', or 'Does using a mobile phone cause brain damage?')
- giving your opinion on a science-related issue that involves balancing risks and benefits, and may also involve considering values and beliefs (for example, 'Should research into

human cloning be allowed?', 'Should the government require all babies to be vaccinated?')
- making personal or social choices (for example, 'Should I ask my parents for a lift to school or should I take the bus?').

Before you make your final choice of topic, it is worth checking these things.
- Will you be able to explain the scientific ideas connected to the study when you write your report?
- Can you find enough information to help you to write your report?

If you don't understand the science very well, or you cannot find much information, you should think about studying a different topic.

Finding and evaluating information

Your teacher may provide you with some information to start you off on your case study. However you will also need to find some information yourself, to find out about the different views that people may have on your topic. You may be able to write off to various organisations for information, but you will probably find a lot of the material you need on the internet.

Many sources of information are biased in some way, that is they support only one side of an argument. When you are choosing sources, you need to think carefully about how the source might be biased, and make sure you balance one source with another that presents the other side of the argument.

For example, an environmental organisation may have a website giving all the reasons against genetic modification of crops, and the website of a company that produces GM seeds might present only the other side of the argument. You need to compare information from both sites to arrive at your conclusion.

You also need to think about whether each source of information is providing scientific information, or if it is just presenting opinions that are not necessarily based on science.

Acknowledging your sources

You must provide a list of all the sources of information you use. If you get information from a website, say whose website it is, and give the address and the date when you obtained the information.

You should set out all your references in the same way. You could set them out like this:
- author or website owner
- book/website address/magazine or newspaper title
- article title (if your reference is from a magazine or website)
- page numbers (if your reference is from a book or magazine)

For example:

References

1) NHS website, http://www.mmrthefacts.nhs.uk/basics/whatismmr.php, What is MMR?

2) Bradfield & Potter, Longman GCSE Biology, pages 10–15

If you are directly quoting from a source of information, you must make this clear by putting the words you have quoted in inverted commas (' '), and saying where the quote came from. Make sure that anything you copy directly is relevant to your argument – you won't get any marks for just copying out large chunks of text.

Presenting your report

Your report does not have to be a straightforward account of what you have found out. You can present your case study in different ways, such as:
- a formal report
- a newspaper or magazine article
- a PowerPoint presentation
- posters for a campaign.

However you choose to present your report, it must:

- be written for a target audience
- be clear and concise
- be organised in a sensible way, with titles and subheadings, and a contents list to help people to find the key parts of it
- include tables, charts and diagrams to present data or help you to explain things
- include a list of the sources you have used.

Getting good grades in exams 1

Planning your revision

The grade you get for your GCSE Biology (or Physics or Chemistry) depends on the marks you get in lots of different assessments. You give yourself the best chance of an A* if you do as well as you can on all the different components of your course. This means that you need to revise before each exam.

When to revise

Find out from your teacher when you will be taking the module exams, and which modules will be tested in each exam. You also need to find out if you will be taking exams in other subjects at the same time, so you can plan all your revision together.

Don't leave it too late! Plan to start revising several weeks before the exam. That way, you won't get stressed by trying to cram all your revision into just a couple of evenings or a weekend.

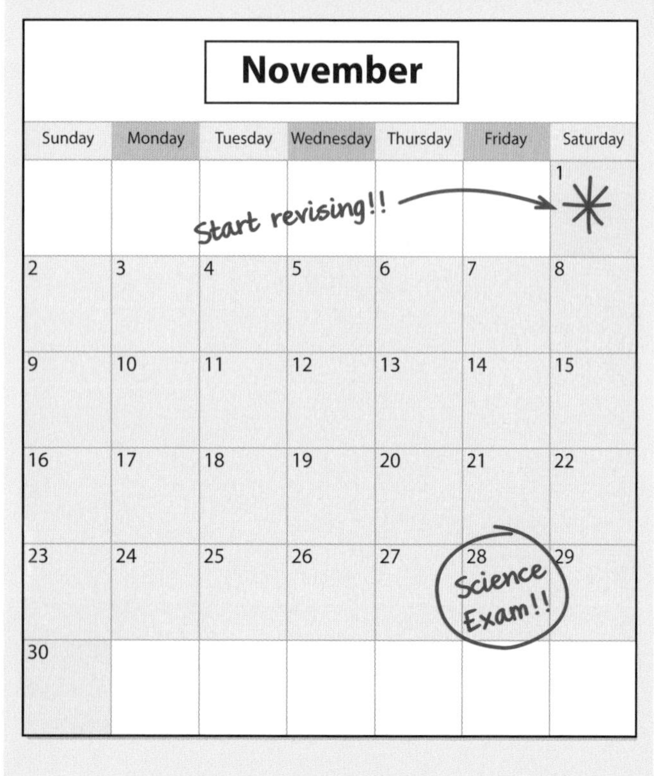

If the exam is part-way through the school year, you won't get study leave to help you to revise. Plan on spending a couple of hours each evening revising, with regular breaks. If you are revising over the weekend, don't try to work right through the weekend. Give yourself an afternoon or an evening off and do something you enjoy – this will help you to relax and you will feel better and more alert when you start revising again.

If you are on study leave, make a timetable for each day and stick to it. Are you a 'morning person' ? If you are, plan on doing most of your revision in the mornings, when you will feel more awake. If you are an evening person, don't try to make yourself work in the mornings – but don't make this an excuse to do less work overall!

Where to revise

You need a quiet place to work, without distractions. Ideally you should be sitting at a table or desk. Lying on your bed with your books will just make you fall asleep, and you cannot watch TV and revise at the same time! Some people find that having music playing helps them to concentrate, but be honest with yourself – does music *really* help you to revise, or is it just a distraction?

You can't revise properly like this!

Making a timetable

Before you draw up your revision timetable:

- Find out what you need to know. This textbook contains all you need to know for each module. You can also ask your teacher for a copy of the Specification, or you can download it from the OCR website.
- Find out how much you already know. Test yourself by doing the summary exercises in this book again, answering the questions at the beginning of each double page, or the pages of questions at the end of each module. Try to answer them without looking at the information on the pages! Make a note of the topics that you found difficult, and plan to spend time on these.

Once you learn a fact it stays in your memory for a while, but it doesn't stay there very long if you only think about it once. To give yourself the best chance of remembering what you learn you need to review what you have just learnt at the beginning of your next session. The diagram shows how this could work.

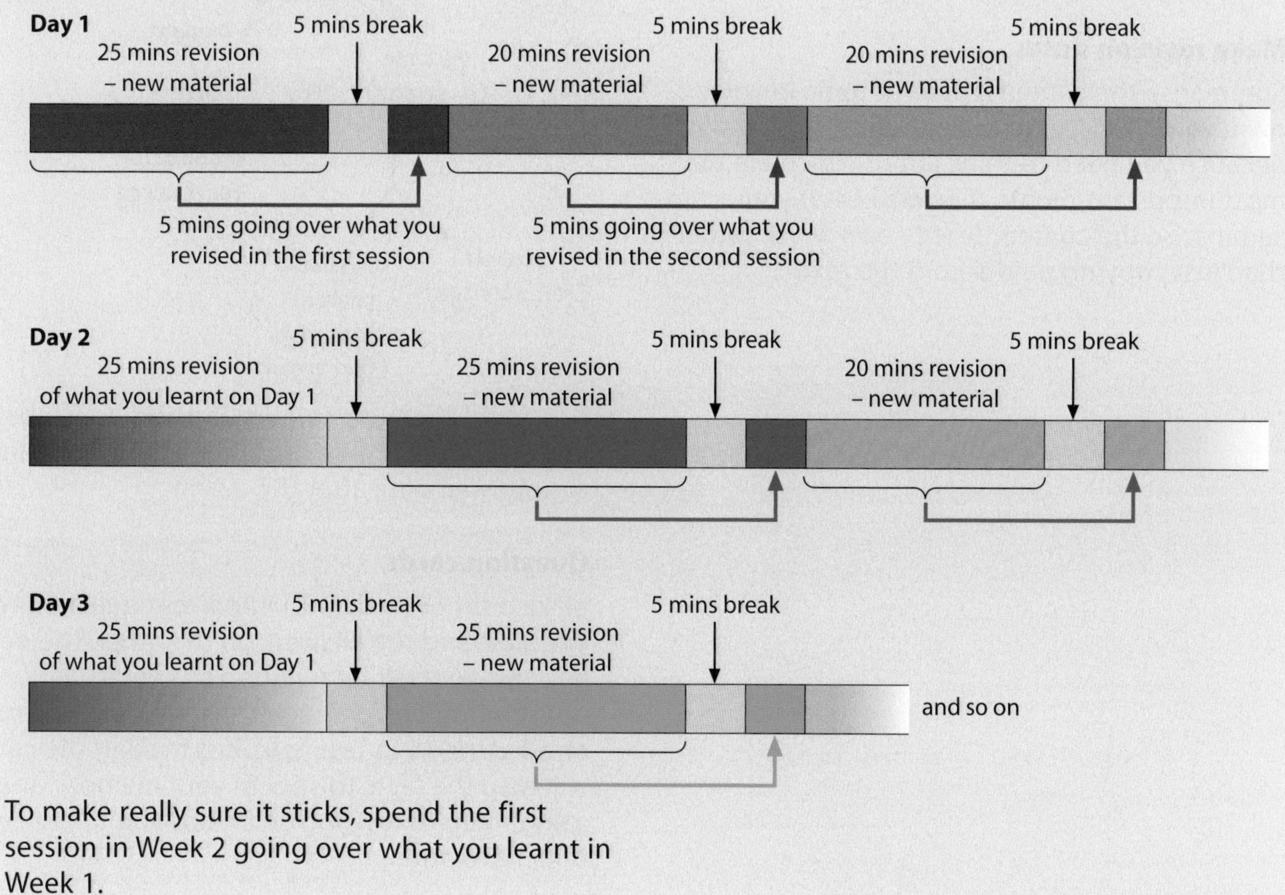

To make really sure it sticks, spend the first session in Week 2 going over what you learnt in Week 1.

Getting good grades in exams 2

How to revise

Revision needs to be *active* – you need to be doing something that makes you think about the things you are trying to learn. Just sitting and reading through your textbook or your exercise book is usually the least effective way of learning things.

Different people have different learning styles, so what works for your friends may not work so well for you. Here are some ideas that can help you to learn the things you need to remember for the exam. Try out the different ways, and then stick with the ones that work best for you.

You also need to *understand* what you are revising, so make a note of any parts of the course you don't understand and ask your teacher about them.

Make revision notes

Summarise the contents of each topic in your own words, as a set of bullet points. This works because you need to think about which are the most important points. It works best if you summarise the content *in your own words* rather than just copying points from the text.

Mind maps

Make Mind maps to summarise important points. Mind maps are a very visual way of recording important facts and how they are linked. Because you have to pick out the key words and decide how they are linked, you have to think about what you are revising, and this helps it to get into your long-term memory.

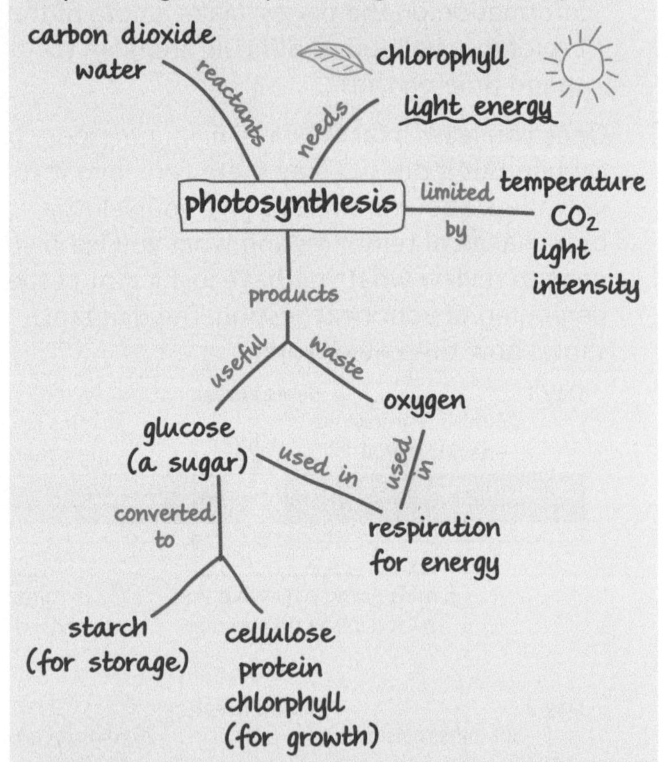

You could even use colours and add drawings to make your mind map look interesting, and pin it to your wall until after the exam!

Question cards

Make a set of cards with important questions on one side, and the answers on the other. The next day, test yourself by trying to answer each question before looking at the back of the card, or ask a friend to test you. Just making the cards will help the facts to stick in your memory, and using them to test yourself helps you to review the information.

Mnemonics, rhymes and raps

These techniques can all help you to learn lists of things. Mnemonics, like the one for the colours of the visible spectrum, are good for remembering lists where the order is important.

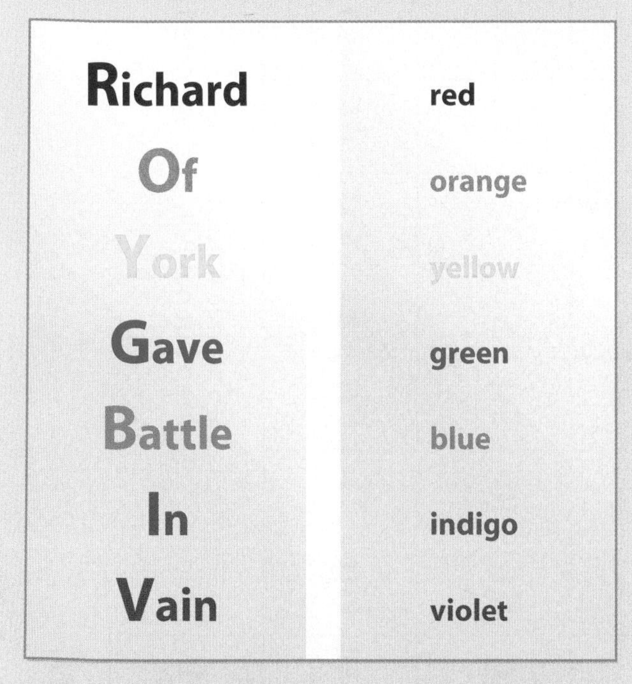

Don't spend ages trying to make up a rhyme or a rap about something (and don't use it as an excuse to waste time when you should be concentrating on science!). If you are good at making up this kind of thing, write one and share it with your friends!

Revise with a friend

Take it in turns to ask each other questions about the topic, and check your answers. But remember you are supposed to be revising, not chatting!

Practice exam questions

Your teacher may be able to give you some past exam papers. Try answering all of the questions in the time allowed, and then use your textbook to help you to check the answers. This will also help you to find out how much more revision you need to do, and give you an idea of how fast you will have to work in the real exam.

How to answer exam questions

You just read the question and write the answer, right? Well, almost.

Some questions just ask you to tick a box or write one or two words, and it is quite easy to understand what the examiner is asking you to do.

Some questions require longer answers, and you need to think about these before starting to write your answer.

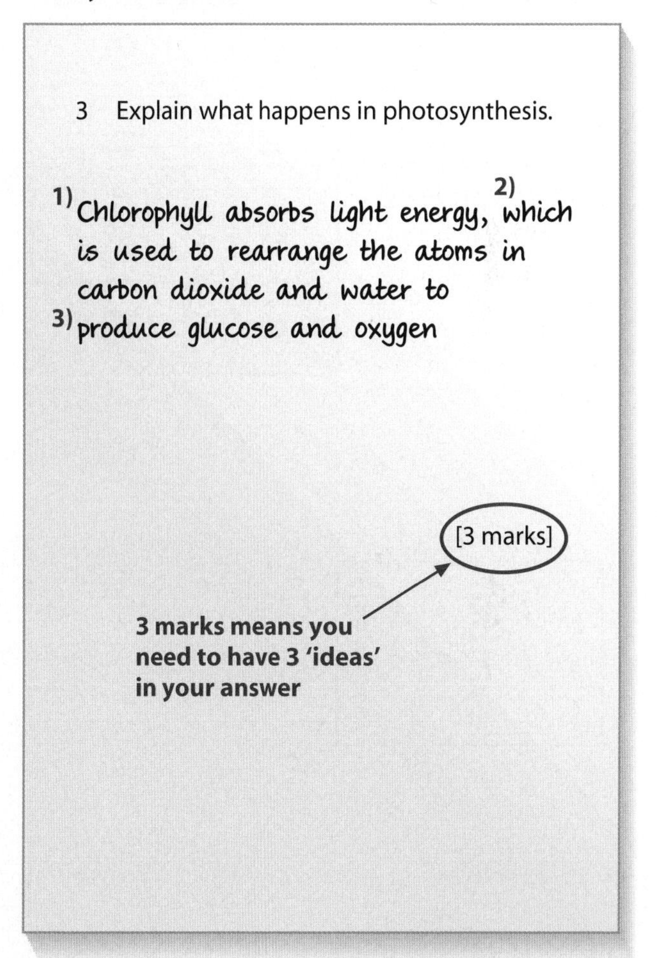

In some cases you might get one of the marks for writing a clear, ordered answer. Take a minute to think about what you are going to write, so you can put your ideas down in a sensible order. Use some of the space on the exam paper to make notes, if you need to, and just cross them out afterwards so the marker knows they are not part of your answer.

Periodic Table

Key

mass number
symbol name
atomic number

1	2											3	4	5	6	7	0
					1 **H** hydrogen 1												4 **He** helium 2
7 **Li** lithium 3	9 **Be** beryllium 4											11 **B** boron 5	12 **C** carbon 6	14 **N** nitrogen 7	16 **O** oxygen 8	19 **F** fluorine 9	20 **Ne** neon 10
23 **Na** sodium 11	24 **Mg** magnesium 12											27 **Al** aluminium 13	28 **Si** silicon 14	31 **P** phosphorus 15	32 **S** sulphur 16	35.5 **Cl** chlorine 17	40 **Ar** argon 18
39 **K** potassium 19	40 **Ca** calcium 20	45 **Sc** scandium 21	48 **Ti** titanium 22	51 **V** vanadium 23	52 **Cr** chromium 24	55 **Mn** manganese 25	56 **Fe** iron 26	59 **Co** cobalt 27	59 **Ni** nickel 28	63.5 **Cu** copper 29	65 **Zn** zinc 30	70 **Ga** gallium 31	73 **Ge** germanium 32	75 **As** arsenic 33	79 **Se** selenium 34	80 **Br** bromine 35	84 **Kr** krypton 36
85 **Rb** rubidium 37	88 **Sr** strontium 38	89 **Y** yttrium 39	91 **Zr** zirconium 40	93 **Nb** niobium 41	96 **Mo** molybdenum 42	[98] **Tc** technetium 43	101 **Ru** ruthenium 44	103 **Rh** rhodium 45	106 **Pd** palladium 46	108 **Ag** silver 47	112 **Cd** cadmium 48	115 **In** indium 49	119 **Sn** tin 50	122 **Sb** antimony 51	128 **Te** tellurium 52	127 **I** iodine 53	131 **Xe** xenon 54
133 **Cs** caesium 55	137 **Ba** barium 56	139 **La*** lanthanum 57	178 **Hf** hafnium 72	181 **Ta** tantalum 73	184 **W** tungsten 74	186 **Re** rhenium 75	190 **Os** osmium 76	192 **Ir** iridium 77	195 **Pt** platinum 78	197 **Au** gold 79	201 **Hg** mercury 80	204 **Tl** thallium 81	207 **Pb** lead 82	209 **Bi** bismuth 83	[209] **Po** polonium 84	[210] **At** astatine 85	[222] **Rn** radon 86
[223] **Fr** francium 87	[226] **Ra** radium 88	[227] **Ac*** actinium 89	[261] **Rf** rutherfordium 104	[262] **Db** dubnium 105	[266] **Sg** seaborgium 106	[264] **Bh** bohrium 107	[277] **Hs** hassium 108	[268] **Mt** meitnerium 109	[271] **Ds** darmstadtium 110	[272] **Rg** roentgenium 111							

Elements with atomic numbers 112–116 have been reported but not fully authenticated.

* The Lanthanides (atomic numbers 58–71) and the Actinides (atomic numbers 90–103) have been omitted.
The mass numbers for Cu and Cl have not been rounded to the nearest whole number.

Glossary

absolute zero Lowest possible temperature. All movement stops at absolute zero.

absorption spectrum Light from a star spread out into a spectrum, with black lines across it where certain elements have absorbed different wavelengths of light.

activation energy The amount of energy needed to break bonds at the start of a reaction.

H **active transport** Using energy to pump molecules in the opposite direction to which they would normally go by diffusion.

acute Sudden.

aerobic exercise Exercise that uses aerobic respiration and increases heart rate and breathing rate.

aerobic respiration Chemical reaction in which glucose is reacted with oxygen to produce carbon dioxide and water, releasing energy.

alcohol An organic compound containing an –OH functional group.

alkane A hydrocarbon containing only carbon to carbon single bonds.

alkene A hydrocarbon containing carbon to carbon double bonds.

H **allele** Different forms of a gene controlling the same characteristic (e.g. a flower colour gene may come in two alleles, one for red and one for white).

altitude (astronomy) The angle above the horizon to look for a particular star.

amino acid Building block of proteins.

anaerobic respiration The release of energy from food molecules without using oxygen.

anaerobic Lacking oxygen.

angular magnification How much the image of an object is magnified. Calculated using the focal length of the objective divided by the focal length of the eyepiece.

anhydrous A substance that contains no extra water molecules.

antagonistic pair Pair of muscles that work together to move a limb.

antibiotic Chemical that can kill bacteria.

antibody Protein that attaches to other proteins (called antigens) on the surfaces of foreign cells to help destroy the cells.

antigen Protein on the surface of a cell.

aorta Major artery leading from the heart to the rest of the body.

aperture The hole through which light enters a telescope. In a simple telescope, this will also be the size of the objective.

apparent brightness How bright a star appears to be when seen from Earth.

aqueous solvent A solvent that contains water.

arcsecond (") 1/3600 of a degree

artery Blood vessel taking blood away from the heart.

arthritis Diseases that cause inflammation of joints.

articulate Move, as at a joint.

atom economy The actual amount of product formed as a percentage of what should be formed.

H **ATP** The small molecule that is formed using energy released in respiration. (ATP stands for adenosine triphosphate.)

atrium Upper chamber of the heart, which receives blood from veins.

autoradiography Using a radioactive substance to mark a gene probe so that it can be found again using photographic paper or X-ray film.

autotroph An organism that makes its food (also called a producer). Green plants are autotrophs.

azimuth (astronomy) The horizontal direction in which to look for a star, measured as an angle from north.

backward reaction Where the products recombine to form the reactants in an equilibrium reaction.

H **balanced symbol equation** A way of representing chemical reactions in which the number of reactant atoms equals the number of product atoms.

ball and stick model A way of showing what is in a molecule by drawing the atoms as balls and the bonds as sticks.

H **baseline** Assessment done at start of treatment.

biomass The mass of living and dead organisms.

black hole Core of a red supergiant that has collapsed. Black holes are formed if the remaining core has a mass more than three or four times that of the Sun.

blood pressure The pressure of blood measured in the arteries.

blood type A, B, AB or O are the common blood types, and show the antigens on red blood cells.

blood vessel Tube that carries blood around the body.

H **bond energy** The specific amount of energy needed to break a bond, or that is given out when that bond forms.

bulk chemicals Chemicals made by the chemical industry on a very large scale.

burette Apparatus used in titrations to measure out accurate volumes of liquids.

by-products Unwanted products of a chemical reaction.

calibration curve A graph of colour intensity against concentration, used to find out the concentration of a solute in solution.

capillary bed Network of fine capillaries in a tissue.

capillary Narrow, thin-walled blood vessel running through tissues.

carboxylic acid An organic compound with a –COOH functional group.

H **carrier** Someone who has a recessive allele that can be passed on to his/her children but who does not have the disease that this allele causes.

cartilage Smooth tissue that covers the ends of bones in joints.

catalyst A substance that speeds up a reaction without being used up.

celestial equator An imaginary line in the sky above the Earth's equator.

celestial latitude The position of a star in space measured as an angle upwards from the celestial equator. Also called declination.

celestial longitude The position of a star in space measured as an angle along the celestial equator, starting from a point called the vernal equinox. Also called right ascension.

celestial poles Imaginary points in space above the Earth's north and south poles.

cell membrane Outer part of a cell that controls what goes into and out of the cell.

cell wall A part of plant and bacterial cells that protects the cell and holds it together. Bacterial cell walls are soft.

cellulose Polymer of glucose used in plant cell walls.

Cepheid variable A variable star whose brightness varies with a regular period, where the intrinsic brightness is related to the period.

chamber Compartment of the heart.

chlorophyll Green pigment found in chloroplasts that absorbs light energy to power photosynthesis.

chloroplast Disc-shaped structure found in plant cells that contains chlorophyll.

chromosome Large molecule of DNA that controls the production of proteins.

chronic Happens over a long time.

circulatory system The heart and blood vessels, which take blood around the body.

H **codominant** Two alleles of the same gene that both have effects when they are found together in a cell are codominant.

commensalism A relationship between two organisms such that one organism benefits but the other receives neither harm nor benefit.

H **compensation point** Point at which the amount of carbon dioxide released by respiration in a plant is equal to the amount used in photosynthesis.

concave mirror Type of curved mirror that brings rays of light together at a point.

concentration A measure of how much solute is dissolved in a solvent. It is measured as g/dm^3.

consumer An organism that gets it energy from other organisms.

contract To shorten, as in muscles.

convection Movement of particles in a liquid or gas that carries energy to cooler areas.

converge Come together.

converging lens A lens that is fatter in the middle and brings light rays together to a point.

convex lens Another name for a converging lens.

core Centre of a star.

cytoplasm The fluid part of a cell in which many reactions occur.

decay Break down to simpler chemicals.

declination The position of a star in space measured as an angle upwards from the celestial equator. Stars south of the celestial equator have a negative value of declination.

denature When an enzyme loses its shape and stops working it is said to be denatured.

diffraction When rays of light bend round an obstacle or when they spread out as they pass through a small gap.

diffusion The net movement of particles from an area in which they are in high concentration to an area in which they are in lower concentration.

dislocation Movement of bone out of position in a joint.

dissolving quantitatively Accurately dissolving a solute in a solvent so that the concentration can be found.

H **dominant** An allele that prevents another allele of the same gene from having an effect is dominant.

double circulatory system A circulatory system with two loops, one going to the lungs and the other to the rest of the body.

drugs Substances that affect chemical reactions in our bodies.

dynamic equilibrium Where the forward reaction and backwards reaction occur at the same time and at the same rate. The amounts of each reactant present do not change.

ecosystem A group of plants and animals, and the physical environment in which they live, such as a pond or rainforest.

efficiency of energy transfer The proportion of energy in one trophic level that is converted to body tissue in the next.

egest Remove from the body without being absorbed. Faeces are egested.

endothermic reaction A reaction that takes in heat energy from the surroundings.

end-point Where the reaction is complete during a titration.

energy level diagram (chemistry) A diagram that shows how energy changes during an endothermic or exothermic reaction.

energy level diagram (physics) A diagram showing the different amounts of energy that electrons in an atom can have.

enzyme A protein that speeds up reactions in organisms.

equatorial system The way of referring to the position of a star using celestial latitude and longitude or declination and right ascension.

equilibrium Where reactants form products, which can then react together again to form the reactants.

ester An organic compound formed by the reaction between an alcohol and a carboxylic acid.

exothermic reaction A reaction that gives out heat energy to the surroundings.

eyepiece The lens that you look through in a simple telescope.

fatty acid A long-chained carboxylic acid.

feedstocks The starting materials of a reaction.

fermentation Process when microorganisms are grown on a large scale to obtain a useful product.

fine chemicals Pure chemicals made in small amounts by the chemical industry.

fluorescent Something which glows when ultraviolet light is shone at it.

focal length Distance from the centre of the lens to the focal point.

focal point The point at which parallel rays of light come together when they have passed through a converging lens.

food additives Chemicals added to food during its production.

food chain A diagram that shows the order that organisms feed on each other. It also shows the transfer of energy between the trophic levels.

forward reaction The reaction in which the reactants form the products in an equilibrium reaction.

fracture Breaking of bone.

fragrances Mixes of chemicals with specific smells.

functional group A group of atoms that give an organic compound its properties.

fusion When two nuclei join together and release energy.

galaxy A group of millions of stars held together by gravity.

gas-liquid chromatography Chromatography used to separate tiny amounts of substances, using a very long thin column.

gene probe Short section of DNA used to find a whole gene.

gene transfer Getting a foreign gene into an organism so that it will work in that new organism.

gene Part of a chromosome. One gene contains the instructions for making a particular protein.

genetic modification Altering the genes of an organism to make it do something differently.

genetic test Testing people for alleles that can cause diseases.

glucose A type of sugar ($C_6H_{12}O_6$) made in photosynthesis.

glycerol The alcohol that forms an ester in fats and oils.

green chemistry Principles for sustainable chemistry with a minimum of waste and toxic chemicals.

haemoglobin Protein found in red blood cells that is able to bind to oxygen molecules to it.

heart rate The number of times each minute that the heart contracts.

heart Muscular pump in the body.

heterotroph An organism that gets its energy from other organisms.

host The organism that a parasite feeds on.

Hubble constant A number used to relate the speed that a galaxy is moving away from us to its distance from us. It is used in the equation: speed (km/s) = Hubble constant (km/s/Mpc) [mult] distance (pc).

hydrated A substance that contains water molecules as part of its structure.

hydration A reaction with water or steam.

hydrocarbon A compound that contains only hydrogen and carbon atoms.

image A 'picture' of something produced by a lens.

indicator A chemical that changes colour at the end-point of a titration.

inflame To swell.

inorganic material Material that does not come from living things.

intrinsic brightness The true brightness of a star, if viewed from nearby.

inverted Upside down.

ionise To lose electrons.

isolation Finding the gene that causes a useful characteristic in an organism.

Kelvin temperature scale A temperature scale with units the same size as degrees Celsius, but with absolute zero as its starting point.

lactic acid Product of anaerobic respiration.

leach Removal of substances, e.g. mineral ions, in water draining through soil.

ligament Tough tissue that joins bone to bone.

light pollution Light radiated into space from street lights, buildings, etc.

light year The distance light travels in a year. It is approximately 9.467 [mult] 10^{15} m.

limiting factor A factor (e.g. temperature, light, amount of carbon dioxide) that slows the speed of a reaction when there is not enough of it.

locating agent A chemical used to show the presence of colourless spots during chromatography.

lunar eclipse When the Moon goes into the shadow of the Earth.

megaparsec (Mpc) 1 million parsecs.

mobile phase The solvent that moves through the stationary phase during chromatography.

molecular formula The formula that shows how many atoms of each element there are in a molecule.

H **natural selection** When certain variations in characteristics allow some organisms to survive and breed in an area, whilst others with less suitable variations do not survive.

nebula A cloud of gas in space. Some objects that look like nebulae are actually other galaxies or clusters of stars.

neutron star Core of a red supergiant that has collapsed. All of the particles become neutrons if the remaining core has a mass between about 1.5 times and three or four times that of the Sun.

nitrate Type of mineral ion needed for making proteins in plants.

non-aqueous solvent A solvent that does not contain any water.

nucleus Centre of an atom that contains protons and neutrons.

objective The lens in a simple telescope that is nearest to the thing you are looking at.

optimum The conditions at which a reaction happens fastest.

organic compound A compound that contains carbon (except for metal carbonates, carbon dioxide and carbon monoxide).

organic material Material that comes from living things.

origin Where the spots are placed on a chromatogram.

H **osmosis** The net flow of water molecules from where there are more of them (a dilute solution) to where there are less, through a partially permeable membrane.

H **osmotic balance** When there is no net flow of water molecules from one side of a partially permeable membrane to the other.

H **oxygen debt** Oxygen needed after anaerobic respiration to break down lactic acid.

oxyhaemoglobin Haemoglobin with oxygen attached to it.

paper chromatography Chromatography where paper is the stationary phase.

parallax angle Half of the angle that a star appears to move over 6 months, against the background of distant stars.

parallax The apparent movement over 6 months of nearby stars against the background of distant stars, due to viewing the star from different positions as the Earth moves around its orbit.

parasite An organism that lives in close association with another organism (the host) to obtain nutrition, and which causes damage to the host.

parasitism A feeding relationship between two organisms such that one organism benefits and the other is harmed.

parsec The distance of a star whose parallax angle is 1 arcsecond. It is approximately 3.068 [mult] 10^{16} m.

partial eclipse When the Sun is only partly blocked out by the Moon (for a solar eclipse) or when only part of the Moon goes into the shadow of the Earth (for a lunar eclipse).

H **partially permeable membrane** A thin sheet with tiny holes that allow small molecules through (e.g. water) but are too small to let larger molecules through.

period A length of time. For a Cepheid variable, it is the length of time from one maximum brightness to the next.

phases of the Moon The different shapes that the Moon appears to have at different times.

photosphere Surface of a star, where energy is radiated into space.

photosynthesis A set of chemical reactions in which water and carbon dioxide are turned into glucose and oxygen using energy from light.

physiotherapist A trained medical person who treats injuries with exercises, massage, etc.

pipette Apparatus used to measure a certain volume of liquid accurately.

planetary nebula Material that is ejected from the outer layers of a red giant when it collapses to form a white dwarf. (It is not connected with planets.)

plasma The liquid component of blood that carries dissolved solutes and proteins, such as antibodies.

H **plasmid** Small circle of DNA found in the cytoplasm of bacteria containing genes for a small number of proteins.

platelet Tiny cell fragment that helps blood to clot at sites of injury.

polymer Many molecules of one type of chemical linked together to form a long chain.

power How strong a lens is. Calculated from 1/focal length (in metres). The unit for power is the dioptre.

pressure Force produced by collisions of molecules of a gas with each other and their container.

primary mirror The first, curved, mirror that rays of light hit in a telescope. Used in place of an objective lens.

producer An organism that makes its food, i.e. an autotroph.

protein Polymer of different amino acids.

protostar Cloud of gas drawn together by gravity which has not yet started producing its own energy.

H **protozoan** A type of single-celled organism.

pulmonary artery Blood vessel taking deoxygenated blood from the heart to the lungs.

pulmonary vein Blood vessel taking oxygenated blood from the lungs to the heart.

pulse The feel of your blood being pumped through your arteries.

H **Punnett square** A square used to work out the chance of offspring having a certain combination of alleles.

pyramid of biomass A diagram that shows the mass of living material at each trophic level in a food chain.

pyramid of numbers A diagram that shows the numbers of organisms at each trophic level in a food chain.

qualitative analysis A method of finding out what is in a substance.

quantitative analysis A method of finding out how much of a particular substance there is in a sample.

H **RAM** Another way of writing relative atomic mass.

rate The speed of a reaction is known as its rate.

raw materials The substances needed in order to make another substance in a chemical reaction.

real (physics) An image formed when light rays converge. A real image can be formed on a screen or piece of paper.

H recessive An allele that is prevented from having an effect by another allele is recessive.

red blood cell Cell without a nucleus, packed with haemoglobin.

red giant A star that has used up all the hydrogen in its core and is now fusing helium to make bigger atoms.

red supergiant Star that has used up the hydrogen in its core and has a mass much higher than that of the Sun.

redshift The movement of absorption lines towards the red end of the spectrum caused by a star or galaxy moving away from us.

H reflux A means of heating a liquid so that the liquid evaporates and re-condenses so it is not lost from the reaction mixture.

refraction Change in speed when light goes from one transparent material to another. This often causes a change in direction.

rehabilitation Exercises and other treatment, after injury, to recover as much normal movement as possible.

H relative atomic mass The mass of an atom of an element compared with the mass of one hydrogen atom.

H relative formula mass The mass of a molecule on the same scale as relative atomic mass.

rennin Enzyme used to turn parts of milk into solid lumps, which are then used to make cheese.

replication Making many copies of a gene.

retention factor The ratio of distance travelled by a spot to the distance travelled by the solvent front, used to identify substances separated by chromatography.

retention time How long it takes for a sample to pass through a gas-liquid chromatography column

retrograde motion When a planet appears to move backwards against the background of distant stars.

reversible reaction A reaction in which the products can be made to react together to form the reactants again.

H RFM Short for relative formula mass.

RICE Rest, Ice, Compression, Elevation: the first-aid routine used to treat skeleto-muscular injuries.

right ascension (RA) The position of a star in space measured as an angle along the celestial equator, starting from a point called the vernal equinox. It can be given in hours or in degrees.

risk factor Factor that is linked to increased chance of a disease or injury.

H root hair cell A root cell adapted to absorbing water quickly by having a projection called a root hair, which increases the surface area.

saturated Contains only carbon–carbon single bonds.

secondary mirror The mirror in a reflecting telescope that deflects the rays of light towards the eyepiece lens.

H sickle-cell anaemia Inherited disease in which red blood cells become sickle-shaped when they lose oxygen and can clump together, blocking blood vessels.

side-effect A negative effect of taking a drug, such as headache.

sidereal day A day measured using the stars. It is shorter than a solar day.

single-cell protein Protein-rich food that is obtained from microorganisms.

solar day A day measured using the Sun.

solar eclipse When the Moon is between the Sun and the Earth and casts a shadow on part of the Earth.

solute Something that dissolves in a solvent to form a solution.

solvent front The distance travelled by the solvent from the origin during chromatography.

solvent A liquid in which solutes dissolve.

speciality chemicals Another word for fine chemicals.

sprain Injury to a ligament.

standard flask A glass flask used to make up solutions of known concentrations.

standard reference materials Known substances used to help identify unknown substances during chromatography.

starch A carbohydrate made from glucose molecules joined together. Plants use it as a store of energy.

stationary phase The solid through which the solvent passes during chromatography.

stomata Holes in leaf surfaces, through which gases can diffuse. Singular = stoma.

strong acid An acid which is totally ionised in water.

strong nuclear force The force that holds protons and neutrons together in the nucleus.

structural formula A way of showing how atoms are grouped together in a molecule.

sugar A type of soluble carbohydrate.

supernova Explosion produced when the core of a red supergiant has become mainly iron and collapses.

symbiosis A relationship between two organisms such that both organisms benefit.

synovial capsule Capsule of tough tissue surrounding a joint.

synovial fluid Viscous fluid that fills a synovial capsule, acting as lubricant and providing nutrients to cartilage.

synthesis Making a new chemical from simple starting materials.

telescope Instrument used to look at distant objects in more detail.

tendon Tough tissue that joins muscles to bones.

thin-layer chromatography Chromatography where a thin layer of alumina or silica is the stationary phase.

tissue fluid Fluid containing oxygen, glucose, etc. surrounding the cells in tissues.

titration A method of accurately finding the concentration of solutions.

total eclipse When the Moon completely blocks out the Sun (for a solar eclipse) or when the Moon is completely in the shadow of the Earth (for a lunar eclipse).

triglyceride The ester formed from glycerol and fatty acids in a fat or oil.

trophic level One level of a food chain, such as producer.

unsaturated Contains at least one carbon–carbon double bond.

urea Waste product produced by the reactions in cells.

valve Flap of tissue that stops blood flowing in the wrong direction in the heart and veins.

variable star A star whose brightness changes.

H **vector** Something that can carry a gene into an organism.

vein Blood vessel taking blood to the heart.

vena cava Major vein leading into the heart.

ventricle Lower chamber of the heart, which pushes blood into arteries leading away from the heart.

vernal equinox A point in space used as the start point for measuring right ascension or celestial longitude.

volatile Easily turned into a vapour.

weak acid An acid which is only partially ionised in water.

white blood cell Blood cell that helps destroy microorganisms in the body.

white dwarf Very dense star that is not very bright. A red giant turns into a white dwarf.

xylem tissue Dead cells in a plant that form long tubes through which water is transported.

yeast A single-celled fungus used to make bread and alcoholic drinks.

Index

Licence Agreement: *21st Century Science GCSE Further Science Units CD-ROM*

Warning:
This is a legally binding agreement between You (the user) and Pearson Education Limited of Edinburgh Gate, Harlow, Essex, CM20 2JE, United Kingdom ('PEL').

By retaining this Licence, any software media or accompanying written materials or carrying out any of the permitted activities You are agreeing to be bound by the terms and conditions of this Licence. If You do not agree to the terms and conditions of this Licence, do not continue to use the Disk and promptly return the entire publication (this Licence and all software, written materials, packaging and any other component received with it) with Your sales receipt to Your supplier for a full refund.

21st Century Science GCSE Further Science Units CD-ROM consists of copyright software and data. The copyright is owned by PEL. You only own the disk on which the software is supplied. If You do not continue to do only what You are allowed to do as contained in this Licence you will be in breach of the Licence and PEL shall have the right to terminate this Licence by written notice and take action to recover from you any damages suffered by PEL as a result of your breach.

Yes, You can:

1. use *21st Century Science GCSE Further Science Units CD-ROM* on your own personal computer as a single individual user;

No, You cannot:

1. copy *21st Century Science GCSE Further Science Units CD-ROM* (other than making one copy for back-up purposes);
2. alter *21st Century Science GCSE Further Science Units CD-ROM*, or in any way reverse engineer, decompile or create a derivative product from the contents of the database or any software included in it;
3. include any software data from *21st Century Science GCSE Further Science Units CD-ROM* in any other product or software materials;
4. rent, hire, lend or sell *21st Century Science GCSE Further Science Units CD-ROM*;
5. copy any part of the documentation except where specifically indicated otherwise;
6. use the software in any way not specified above without the prior written consent of PEL.

Grant of Licence:
PEL grants You, provided You only do what is allowed under the Yes, You can table above, and do nothing under the No, You cannot table above, a non-exclusive, non-transferable Licence to use *21st Century Science GCSE Further Science Units CD-ROM*.

The above terms and conditions of this Licence become operative when using *21st Century Science GCSE Further Science Units CD-ROM*.

Limited Warranty:
PEL warrants that the disk or CD-ROM on which the software is supplied is free from defects in material and workmanship in normal use for ninety (90) days from the date You receive it. This warranty is limited to You and is not transferable.

This limited warranty is void if any damage has resulted from accident, abuse, misapplication, service or modification by someone other than PEL. In no event shall PEL be liable for any damages whatsoever arising out of installation of the software, even if advised of the possibility of such damages. PEL will not be liable for any loss or damage of any nature suffered by any party as a result of reliance upon or reproduction of any errors in the content of the publication.

PEL does not warrant that the functions of the software meet Your requirements or that the media is compatible with any computer system on which it is used or that the operation of the software will be unlimited or error free. You assume responsibility for selecting the software to achieve Your intended results and for the installation of, the use of and the results obtained from the software.

PEL shall not be liable for any loss or damage of any kind (except for personal injury or death) arising from the use of *21st Century Science GCSE Further Science Units CD-ROM* or from errors, deficiencies or faults therein, whether such loss or damage is caused by negligence or otherwise.

The entire liability of PEL and your only remedy shall be replacement free of charge of the components that do not meet this warranty.

No information or advice (oral, written or otherwise) given by PEL or PEL's agents shall create a warranty or in any way increase the scope of this warranty.

To the extent the law permits, PEL disclaims all other warranties, either express or implied, including by way of example and not limitation, warranties of merchantability and fitness for a particular purpose in respect of *21st Century Science GCSE Further Science Units CD-ROM*.

Governing Law:
This Licence will be governed and construed in accordance with English law.